HUMAN GROWTH HORMONE PHARMACOLOGY

Basic and Clinical Aspects

Pharmacology and Toxicology: Basic and Clinical Aspects

Mannfred A. Hollinger, Series Editor
University of California, Davis

Published Titles

Inflammatory Cells and Mediators in Bronchial Asthma, 1990, Devendra K. Agrawal and Robert G. Townley
Pharmacology of the Skin, 1991, Hasan Mukhtar
In Vitro *Methods of Toxicology*, 1992, Ronald R. Watson
Basis of Toxicity Testing, 1992, Donald J. Ecobichon
Human Drug Metabolism from Molecular Biology to Man, 1992, Elizabeth Jeffreys
Platelet Activating Factor Receptor: Signal Mechanisms and Molecular Biology, 1992, Shivendra D. Shukla
Biopharmaceutics of Ocular Drug Delivery, 1992, Peter Edman
Beneficial and Toxic Effects of Aspirin, 1993, Susan E. Feinman
Preclinical and Clinical Modulation of Anticancer Drugs, 1993, Kenneth D. Tew, Peter Houghton, and Janet Houghton
Peroxisome Proliferators: Unique Inducers of Drug-Metabolizing Enzymes, 1994, David E. Moody
Angiotensin II Receptors, Volume I: Molecular Biology, Biochemistry, Pharmacology, and Clinical Perspectives, 1994, Robert R. Ruffolo, Jr.
Angiotensin II Receptors, Volume II: Medicinal Chemistry, 1994, Robert R. Ruffolo, Jr.
Chemical and Structural Approaches to Rational Drug Design, 1994, David B. Weiner and William V. Williams
Biological Approaches to Rational Drug Design, 1994, David B. Weiner and William V. Williams
Direct Allosteric Control of Glutamate Receptors, 1994, M. Palfreyman, I. Reynolds, and P. Skolnick
Genomic and Non-Genomic Effects of Aldosterone, 1994, Martin Wehling

Pharmacology and Toxicology: Basic and Clinical Aspects

Mannfred A. Hollinger, Series Editor
University of California, Davis

Forthcoming Titles

Alcohol Consumption, Cancer and Birth Defects: Mechanisms Involved in Increased Risk Associated with Drinking, Anthony J. Garro
Alternative Methodologies for the Safety Evaluation of Chemicals in the Cosmetic Industry, Nicola Loprieno
Animal Models of Mucosal Inflammation, Timothy S. Gaginella
Antibody Therapeutics, William J. Harris and John R. Adair
Brain Mechanisms and Psychotropic Drugs, A. Baskys and G. Remington
Chemoattractant Ligands and Their Receptors, Richard Horuk
CNS Injuries: Cellular Responses and Pharmacological Strategies, Martin Berry and Ann Logan
Development of Neurotransmitter Regulation and Function: A Pharmacological Approach, Christopher A. Shaw
Drug Delivery Systems, V. V. Ranade
Endothelin Receptors: From Gene to Human, Robert R. Ruffolo, Jr.
Handbook of Mammalian Models in Biomedical Research, David B. Jack
Handbook of Methods in Gastrointestinal Pharmacology, Timothy S. Gaginella
Handbook of Pharmacokinetic and Pharmacodynamic Correlation, Hartmut Derendorf
Handbook of Pharmacology of Aging, 2nd Edition, Jay Roberts
Handbook of Theoretical Models in Biomedical Research, David B. Jack
Immunopharmaceuticals, Edward S. Kimball
Muscarinic Receptor Subtypes in Smooth Muscle, Richard M. Eglen
Neural Control of Airways, Peter J. Barnes
Pharmacological Effects of Ethanol on the Nervous System, Richard A. Deitrich
Pharmacological Regulation of Gene Expression in the CNS, Kalpana Merchant
Pharmacology in Exercise and Sport, Satu M. Somani
Pharmacology of Intestinal Secretion, Timothy S. Gaginella
Phospholipase A2 in Clinical Inflammation: Endogenous Regulation and Pathophysiological Actions, Keith B. Glaser and Peter Vadas
Placental Pharmacology, B. V. Rama Sastry
Placental Toxicology, B. V. Rama Sastry
Receptor Characterization and Regulation, Devendra K. Agrawal
Ryanodine Receptors, Vincenzo Sorrentino
Serotonin and Gastrointestinal Function, Timothy S. Gaginella and James J. Galligan
Stealth Liposomes, D. D. Lasic and F. J. Martin
Targeted Delivery of Imaging Agents, Vladimir P. Torchilin
TAXOL®: Science and Applications, Matthew Suffness
Therapeutic Modulation of Cytokines, M.W. Bodner and Brian Henderson

HUMAN GROWTH HORMONE PHARMACOLOGY

Basic and Clinical Aspects

Edited by

Kathleen T. Shiverick

Arlan L. Rosenbloom

CRC

CRC Press

Boca Raton London Tokyo

Bat S

Library of Congress Cataloging-in-Publication Data

Human growth hormone pharmacology: basic and clinical aspects/edited by Kathleen T. Shiverick and Arlan L. Rosenbloom.
p. 00 cm.—(Pharmacology and toxicology)
Includes bibliographical references and indexes.
ISBN 0-8493-8384-6
1. Somatotropin—Therapeutic use. 2. Somatotropin—Physiological effect.
I. Shiverick, Kathleen T.
II. Rosenbloom, Arlan, L.
III. Series: Pharmacology & toxicology (Boca Raton, Fla.)
[DNLM: 1. Somatotropin—pharmacology. WK 515 H91805 1995]
RM291.2.S6H85 1995
615'.74—dc20
DNLM/DLC
for Library of Congress

94-30842
CIP

Direct all inquiries to CRC Press, Inc., 2000 Corporate Blvd., N.W., Boca Raton, Florida 33431.

International Standard Book Number 0-8493-8384-6
Library of Congress Card Number 94-30842
Printed in the United States of America 1 2 3 4 5 6 7 8 9 0
Printed on acid-free paper

SERIES PREFACE

The series, *Pharmacology and Toxicology: Basic and Clinical Aspects*, has been created in recognition of the fact that, from time to time, a new area of interest within a discipline matures to a critical mass that merits organization and integration of the respective observations into a free-standing monograph. In order for such an undertaking to be successful, each editor and/or author must be qualified to identify and select sources best suited to communicate essential aspects of that subject. Such is the case with *Human Growth Hormone Pharmacology: Basic and Clinical Aspects*. Kathleen T. Shiverick and Arlan L. Rosenbloom have assembled a list of international contributors, whose expertise in their respective areas is well known.

Mannfred A. Hollinger, Ph.D.
Series Editor
Professor
Department of Medical Pharmacology and Toxicology
University of California, Davis
Davis, California

PREFACE

The idea for this volume emerged from a symposium on the clinical and basic aspects of growth hormone organized for the Developmental Pharmacology Section of the American Society for Pharmacology and Experimental Therapeutics in 1993. The goal was to create an up-to-date document in this fast-moving field that would provide basic scientists a resource about clinical aspects and clinical investigators a ready reference on basic and investigative issues. We have been fortunate to be able to recruit the leading investigators in the various areas to contribute chapters.

The breadth of topics addressed, from the most basic molecular to the pragmatics of growth evaluation, is matched by the range of experience reflected in the authorship, covering the entire modern phylogeny of our knowledge of growth and growth hormone from the dawn of the radioimmunoassay era over 30 years ago. Contributing over this entire era has been William Daughaday who provides the closing chapter on the important issue of spontaneous and iatrogenic hypersomatotropism. Other contributions, such as those of Vaccarello and Cohen, provide the promise of the young generation of investigators.

The volume is organized to approximate a textbook approach beginning with factors contributing to GH gene expression in the pituitary and placenta, followed by the description of the GH-prolactin receptor gene family providing insights into functional relationships. Chapter 3, epitomizing state-of-the-art molecular biology from the Genentech group, describes the design of antagonists to the GH receptor, and heralds the future pathway to hormone drug design of antagonists and agonists. Having mastered what is known about the molecular and physiologic aspects of growth hormone, the reader is invited to the next stage in the growth cascade, to understand the insulin-like growth factor axis which requires comprehension of IGFs, their receptors, their binding proteins, and the proteases for the binding protein. Thus, Chapter 4, and the subsequent one on transgenic models to dissect the discrete effects of deficiency or excess of specific growth factors, provide the bridge to the series of clinically oriented discussions. These begin with Chapter 6 on the clinical aspects of growth hormone treatment which outlines the application of GH in conventional and novel circumstances in growing children. The fascinating story of GH insensitivity due to deficiency of the GH receptor epitomizes the recent acceleration of knowledge of the growth cascade and is provided in a comprehensive package including clinical, biochemical, genetic, and treatment aspects. The emergence of the concept of aging as a GH/IGF-I deficiency state is addressed in two chapters which deal with the altered physiology of the somatotropic axis with aging, the relationship of these changes to body composition, and the effects of treatment with GH in both adult GH deficient individuals and the endocrinologically "normal" aged. The immune system as a source and as a responder to GH, GHRH, and IGF-I is a topic bursting with questions, particularly as the therapeutic application of GH or IGF-I is considered for non-deficient states. The final chapter on hypersomatotropism addresses these and other questions about GH excess.

The co-editors have greatly enjoyed assembling and reviewing these chapters and we think that our response reflects the appropriateness of the broad readership this book is designed for, since we are a basic scientist and a clinical endocrinologist/investigator. We hope you share our pleasure in reading this volume.

Finally, we wish to dedicate this book to the memory of Dan Rudman, co-author of Chapter 9, an imaginative and genteel clinical investigator who died shortly after submitting his chapter with Adil A. Abbasi.

THE EDITORS

Kathleen T. Shiverick, Ph.D., is Professor of Pharmacology and Therapeutics, and Adjunct Professor of Obstetrics and Gynecology in the College of Medicine at the University of Florida in Gainesville.

Dr. Shiverick received her B.S. degree from the University of Vermont in 1965. She received her Ph.D. degree from the Department of Physiology, University of Vermont School of Medicine, Burlington, in 1974. Postdoctoral work in developmental endocrinology and pharmacology was done at McGill University and the Royal Victoria Hospital, Montreal, Canada, from 1974 to 1978. She was appointed an Assistant Professor of Pharmacology and Therapeutics at the University of Florida in 1978, became an Associate Professor in 1983, and Professor in 1988.

Dr. Shiverick is a member of the Endocrine Society, American Association for Pharmacology and Experimental Therapeutics, American Society for Immunology of Reproduction, Society of Toxicology, International Society for the Study of Xenobiotics, and the honorary societies Phi Beta Kappa and Phi Kappa Phi.

She has been the recipient of a fellowship from the American Lung Association (1974–1976) and a Fogarty Senior International Fellowship (1992–1995).

Dr. Shiverick has been the recipient of research grants from the National Institutes of Health, the American Heart Association, the American Lung Association, and the Fogarty International Center. She has published more than 60 papers and book chapters. Her current research interests are in the regulation of fetal and placental growth.

Arlan L. Rosenbloom, M.D., is Professor of Pediatrics and Chief of the Division of Pediatric Endocrinology in the College of Medicine at the University of Florida in Gainesville.

Dr. Rosenbloom received his B.A. degree from the University of Wisconsin in Madison in 1955 and his M.D. from Wisconsin in 1958. He served a rotating internship at Los Angeles County General Hospital and was a resident in general practice at General Hospital, Ventura, California. He then spent two years as a physician and surgeon in southeast Asia with MEDICO before returning to a residency in pediatrics and a fellowship in pediatric endocrinology at the University of Wisconsin from 1962 to 1966. He was a medical epidemiologist/advisor in the West African Smallpox Eradication and Measles Control Program based in Yaounde, Cameroon, during 1966 to 1968. He was appointed Assistant Professor of Pediatrics at the University of Florida, Gainesville, in 1968, where he founded the Pediatric Endocrinology Section. He became an Associate Professor in 1971 and Professor in 1974.

Dr. Rosenbloom is a member of the American Academy of Pediatrics, American College of Epidemiology (fellow), American Diabetes Association, American Pediatric Society, Endocrine Society, Florida Medical Association, International Society for Pediatric and Adolescent Diabetes (founding member), Lawson

Wilkins Pediatric Endocrine Society, National Council for International Health, and the Society for Pediatric Research.

He was the recipient of a faculty development award in 1974 and spent the year as Visiting Professor at MacMaster University in Hamilton, Ontario. He has been the recipient of numerous research grants from the National Institutes of Health, the Food and Drug Administration, The National Foundation-March of Dimes, The Robert Wood Johnson Foundation, and private industry. Dr. Rosenbloom has published over 200 papers and 40 books or chapters in books, predominantly dealing with diabetes mellitus and growth.

Dr. Rosenbloom is listed in *Who's Who in America, The Best Doctors in America, American Men and Women of Science* and *Who's Who in Science and Engineering*. He is a frequent guest lecturer at national and international meetings and institutions. He has recently received the faculty research prize in clinical science from the University of Florida College of Medicine and has been named to receive the distinguished alumnus citation for 1995 from the University of Wisconsin Medical Alumni Association.

CONTRIBUTORS

Adil A. Abbasi, M.D.
Department of Medicine
Veterans Administration Center
Medical College of Wisconsin
Milwaukee, Wisconsin

J. Edwin Blalock, Ph.D.
Department of Physiology and Biophysics and Center for Neuroimmunology
University of Alabama at Birmingham
Birmingham, Alabama

Helene Buteau, B.Sc., M.S.
INSERM
Unité 344-Endocrinologie Moléculaire
Faculté de Médecine Necker
Paris, France

Nancy E. Cooke, M.D.
Departments of Medicine and Genetics
University of Pennsylvania
Philadelphia, Pennsylvania

Pinchas Cohen, M.D.
Division of Endocrinology
Department of Pediatrics
University of Pennsylvania
Children's Hospital of Philadelphia
Philadelphia, Pennsylvania

Brian C. Cunningham
Department of Protein Engineering
Genentech, Inc.
South San Francisco, California

William H. Daughaday, M.D.
Division of Metabolism
Department of Internal Medicine
Washington University School of Medicine
Washington University
St. Louis, Missouri

Abraham M. de Vos, Ph.D.
Department of Protein Engineering
Genentech, Inc.
South San Francisco, California

A. Joseph D'Ercole, M.D.
Department of Pediatrics
University of North Carolina at Chapel Hill
Chapel Hill, North Carolina

Helene Dinerstein, B.Sc., M.S.
INSERM
Unité 344-Endocrinologie Moléculaire
Faculté de Médecine Necker
Paris, France

Marc Edery, B.Sc., M.S., Ph.D.
INSERM
Unité 344-Endocrinologie Moléculaire
Faculté de Médecine Necker
Paris, France

Nazario Esposito, B.Sc., M.S.
INSERM
Unité 344-Endocrinologie Moléculaire
Faculté de Médecine Necker
Paris, France

Fatima Ferrag, B.Sc., M.S.
INSERM
Unité 344-Endocrinologie
Moléculaire
Faculté de Médecine Necker
Paris, France

Joëlle Finidori, M.D., M.S., Ph.D.
INSERM
Unité 344-Endocrinologie
Moléculaire
Faculté de Médecine Necker
Paris, France

Germaine Fuh
Department of Protein Engineering
Genentech, Inc.
South San Francisco, California

Laure Goujon, B.Sc., M.S., Ph.D.
INSERM
Unité 344-Endocrinologie
Moléculaire
Faculté de Médecine Necker
Paris, France

Andrew Hoffman, M.D.
Department of Medicine
Stanford University, and
Medical Service
Veterans Affairs Medical Center
Palo Alto, California

Paul A. Kelly, B.Sc., M.S., Ph.D.
INSERM
Unité 344-Endocrinologie
Moléculaire
Faculté de Médecine Necker
Paris, France

Anthony A. Kossiakoff, Ph.D.
Department of Protein Engineering
Genentech, Inc.
South San Francisco, California

Jean-Jacques Lebrun, B.Sc., M.S., Ph.D.
INSERM
Unité 344-Endocrinologie
Moléculaire
Faculté de Médecine Necker
Paris, France

Stephen A. Liebhaber, M.D.
Departments of Medicine and
Genetics, and
The Howard Hughes Medical
Institute
University of Pennsylvania
Philadelphia, Pennsylvania

Robert Marcus
Department of Medicine
Stanford University, and
Geriatrics Research
Education & Clinical Center
Veterans Affairs Medical Center
Palo Alto, California

Makoto Nagano, D.V.M., M.S., Ph.D.
INSERM
Unité 344-Endocrinologie
Moléculaire
Faculté de Médecine Necker
Paris, France

Alain Pezet, B.Sc., M.S.
INSERM
Unité 344-Endocrinologie
Moléculaire
Faculté de Médecine Necker
Paris, France

Marie-Catherine Postel-Vinay, M.D., M.S.
INSERM
Unité 344-Endocrinologie
Moléculaire
Faculté de Médecine Necker
Paris, France

Alan D. Rogol, M.D., Ph.D.
Departments of Pediatrics and
Pharmacology
University of Virginia
Health Sciences Center
Charlottesville, Virginia

Arlan L. Rosenbloom, M.D.
Professor
Division of Pediatric Endocrinology
Department of Pediatrics
University of Florida
Gainesville, Florida

Ron G. Rosenfeld, M.D.
Department of Pediatrics
Oregon Health Sciences University
Doernbecher Children's Hospital
Portland, Oregon

Daniel Rudman, M.D.
Department of Medicine
Medical College of Wisconsin
Veterans Administration Center
Milwaukee, Wisconsin

Kathleen T. Shiverick, Ph.D.
Department of Pharmacology and
Therapeutics
University of Florida
Gainesville, Florida

Athanassia Sotiropoulos, B.Sc., M.S.
INSERM
Unité 344-Endocrinologie
Moléculaire
Faculté de Médecine Necker
Paris, France

Phillipe Touraine, M.D., M.S.
INSERM
Unité 344-Endocrinologie
Moléculaire
Faculté de Médecine Necker
Paris, France

Mark H. Ultsch
Department of Protein Engineering
Genentech, Inc.
South San Francisco, California

Mary A. Vaccarello, M.D.
Departments of Pediatrics and
Pharmacology and Therapeutics
University of Florida
Gainesville, Florida

Douglas A. Weigent, Ph.D.
Department of Physiology and
Biophysics and Center for
Neuroimmunology
University of Alabama at
Birmingham
Birmingham, Alabama

James A. Wells, Ph.D.
Department of Protein Engineering
Genentech, Inc.
South San Francisco, California

CONTENTS

Chapter 1

HUMAN GROWTH HORMONE GENE EXPRESSION IN PITUITARY AND PLACENTA

Nancy E. Cooke and Stephen A. Liebhaber

TABLE OF CONTENTS

I. INTRODUCTION

The growth hormone (GH) gene cluster contains five highly similar genes spanning 48 kb on chromosome 17. The more distantly related prolactin gene is encoded at a single locus on chromosome 6. This organization of the GH-Prl multigene family in humans is distinct from that in subprimate species; rodents and cattle possess a single GH gene and a group of prolactin-like genes.[1-3] The singular composition and organization of the (human) hGH-Prl gene family may reflect recent evolutionary duplication and divergence of a single GH gene precursor.[4] The single hPrl gene is expressed in the pituitary, placenta, and lymphocytes.[3] In contrast, each of the genes in the hGH cluster is expressed either in the pituitary or in the placenta, but not both (Figure 1). Possible additional sites of expression of the hGH genes such as in lymphocytes are incompletely studied at present.

The genes in the GH cluster are all oriented in the same transcriptional direction and their relative positions from 5′ to 3′ are pituitary GH gene (hGH-N), chorionic somatomammotropin-like (hCS-L) gene, hCS-A gene, GH-variant (hGH-V) gene, and the hCS-B gene.[5] The latter four genes are each expressed in the placenta (Figure 1). All five genes in the cluster are comprised of five exons, and each can utilize one or more alternative splicing pathways during RNA processing.[1-3] We have studied some of the actual and putative hormonal isoforms that result from such alternative splicing. The growth hormones, like all peptide growth factors, act through binding to a cognate receptor. The hGH receptor is widely distributed

0-8493-8384-6/95/$0.00+$.50

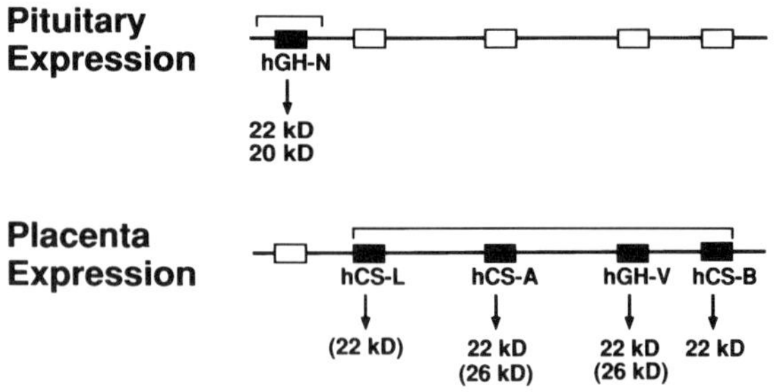

FIGURE 1. The five genes of the hGH gene cluster are shown on each of the two lines. Each rectangle represents a single gene and only those genes active in the indicated tissue are in black; the hGH-N gene is expressed in the pituitary, while the remaining four genes are expressed in the placental villi. The sizes of the encoded proteins are shown in kilodaltons (kD) below each respective gene. Parentheses are used in cases in which the protein product is still hypothetical.

in the body. Of particular interest, we detected an hGH receptor isoform, hGHRd3, in placental villi, and demonstrated that it has the capacity to mediate GH biological actions. The coexpression in the placental villi of a GH hormone, hGH-V, and a cognate receptor, hGHRd3, suggests that the placentally expressed GH gene may act in an autocrine or paracrine manner. Finally, we have begun to study unique forms of gene regulation critical to the transcriptional activation of the cluster in the pituitary by using transgenic mouse models.

II. ALTERNATIVE SPLICING OF THE hGH-N GENE TRANSCRIPT IN THE PITUITARY

The hGH-N gene is expressed as two distinct GH isoforms. Of the hGH in the pituitary and in blood, 90% is 22 kDa and the remaining 10% is 20 kDa. The 20 kDa hGH isoform has a unique spectrum of bioactivity; it exhibits normal growth-promoting action, but its insulin-like activity is reduced to about 20% of the activity of hGH-N.[6] By studying a cell line transfected with the hGH-N gene, we were able to directly demonstrate that the 22 and 20 kDa hGH isoforms are products of the same hGH-N gene and are generated by alternative splicing (Figure 2).[7] Whereas in the majority of transcripts exon 2 splices to the acceptor at the 5′ end of exon 3 (the B acceptor), this alternative splice is to an acceptor site located 45 bases internal to exon 3 (the B′ acceptor). This messenger RNA (mRNA) encodes the 20 kDa variant of hGH-N which contains a 15-amino-acid internal deletion. Surprisingly, although the closely related hGH-V differs from hGH-N by only 13 of 191 amino acids and the hGH-V transcript is identical to the hGH-N transcript for at least 15 nucleotides surrounding both the B and B′ splice-acceptor sites in exon 3, we found that the hGH-V transcript uses only the dominant B acceptor site.

To define the sequences that control the use of B vs. B′ splice-acceptors in the hGH-N transcript we studied the basis for the selective loss of B′ activity in the

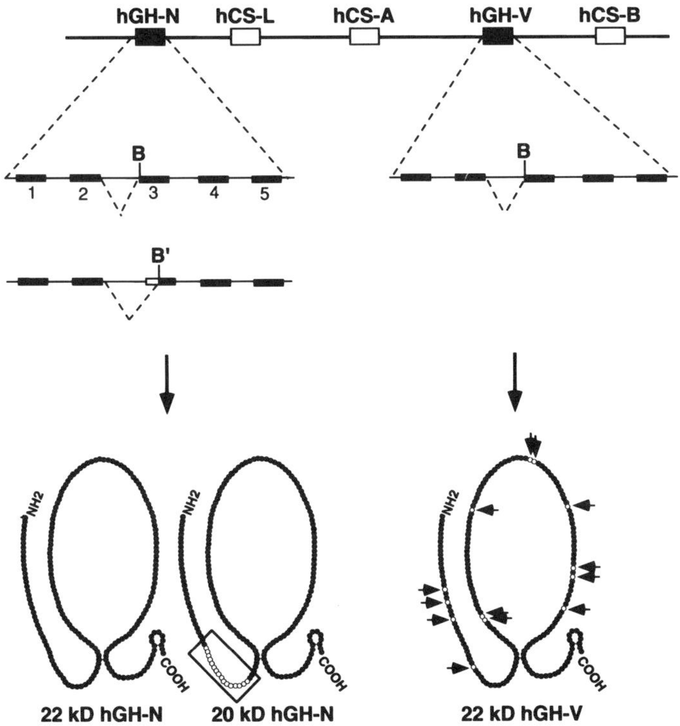

FIGURE 2. Alternative splicing at exon 3 of the hGH genes. The hGH gene cluster is shown on the first line. The exon 2/exon 3 splicing patterns of the hGH-N and hGH-V gene transcripts are shown on the second and third lines. The major (B) and minor (B′) exon 3 splice-acceptor sites are shown. The open segment of exon 3 indicates the region that is excluded from the alternatively spliced hGH-N transcript to generate the 20 kDa hGH-N. Schematics of the three protein products are shown on the last line. Each circle represents a single amino acid residue. The two disulfide bridges are shown. A box marks the 15 residue region of the hGH-N that is deleted in the 20 kDa form. Residues of hGH-V that differ from hGH-N are indicated by the arrows.

hGH-V transcript. This was done by constructing two chimeric genes, hGH-NV3 and hGH-VN3, and studying the splicing pattern of each chimeric transcript after transfection into fibroblasts.[7,8] The hGH-NV3 transcript, in which the third exon and surrounding intron sequences of hGH-N were substituted by those of hGH-V, did not use the B′ site. From this observation we concluded that sequences within the cassetted segment are necessary for usage of the B′ site. We called this *cis* element the ‘‘proximal’’ splicing signal. The reciprocal chimeric gene, hGH-VN3, in which the third exon and surrounding intron sequences of hGH-V were substituted by the corresponding region of the hGH-N transcript, used the B′ acceptor site. This confirmed the localization of the proximal signal within the cassetted exon 3 segment. The level of B′ use in the hGH-VN3 transcript, however, was only 5% rather than the anticipated 10%. We concluded from this observation that the predominant splicing signal resides within the exon 3 cassette

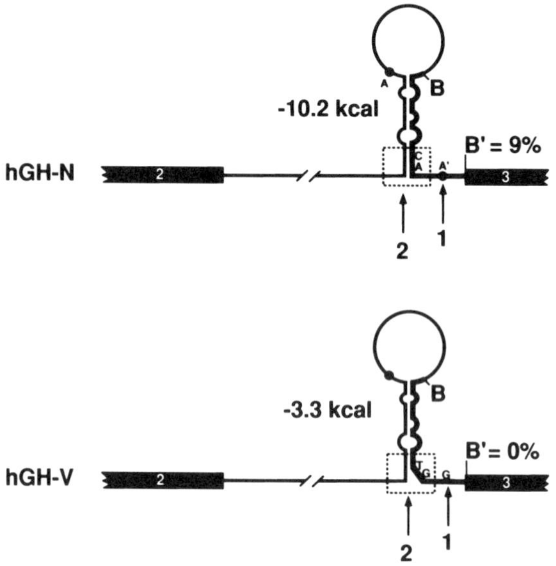

FIGURE 3. The stem-and-loop configuration encompassing the major (B) splice-acceptor site within exon 3 of the hGH-N transcript facilitates utilization of the alternative minor (B′) acceptor. The segment of the hGH-N gene transcript extending from exon 2 through exon 3 is shown. Exons are indicated by rectangles, except the 5′ end of exon 3 which is alternatively spliced and shown as a heavy line. The intron is indicated by a thin line. The proposed secondary structure extends from the 3′ end of intron 2 into the 5′ end of exon 3. The positions of the major (B) and minor (B′) splice-acceptors are shown as are the positions of their two respective lariat-branch points (dots 5′ to the B and B′ sites, respectively). The major lariat-branch point adenosine (A) in the hGH-N transcript is located in the loop. The minor lariat-branch point A (A′) is located 3′ to the stem and is mutated to guanosine (G) in the hGH-V transcript (arrow 1). The two bases, CA, that establish the base of the stable stem of hGH-N transcript are substituted by TG in the hGH-V transcript (arrow 2). The calculated stabilities of the two stems are shown in kilocalories (kcal). The extent of B′ splice-site utilization in each of the two transcripts is noted as a percentage.

(''proximal'' signal), but that an additional ''distal'' signal is located elsewhere in the transcript which is essential for full use of the B′ acceptor site.[7,8]

To dissect the proximal splicing signal, a series of reciprocal and site-directed mutations in the hGH-N and hGH-V genes were generated and each was stably transfected into C127 mouse fibroblasts.[9] Each cell line expressing a particular hGH-N or hGH-V derivative gene was then studied to determine the effects of the specific mutation on B vs. B′ splice-site selection. We identified three bases unique to hGH-N, located between the B and B′ splice-acceptor sites, which are both necessary and sufficient to establish the native splicing pattern of hGH-N when introduced into the hGH-V transcript (Figure 3). One of these bases, an adenosine (A), appears to serve as the lariat-branch point for the B′ acceptor site. This A in the hGH-N transcript is replaced by a guanine (G) in the hGH-V transcript.[8] Two additional intronic bases in the hGH-N transcript are crucial to

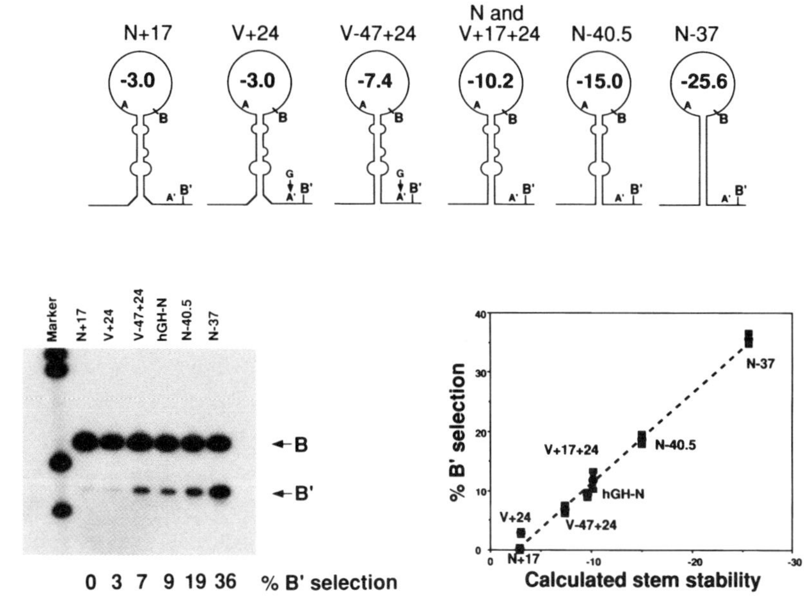

FIGURE 4. Alternative splicing to the B′ acceptor in exon 3 of the hGH-N gene transcripts is directly related to the strength of the stem structure encompassing the major (B) splice-acceptor site. The configurations of the stem and the positioning of the splice-acceptors and their respective lariat-branch point adenosines are as detailed in Figure 3 and in Reference 9. Mutations introduced to alter the stem stability are shown diagrammatically on the first line. The calculated stability of each of the stems is noted within the loop (in kilocalories). The effect of each of these mutations on splicing is determined by an RT/PCR analysis of mRNA isolated from cells stably transfected with each of the mutant genes (for details, see Reference 9). The autoradiograph of the analysis is shown at the lower left and the data are plotted in the graph at the lower right. The data demonstrate a linear relationship between the strength of the predicted stem in each of the mutant transcripts and the degree of alternative splicing. The ordinate represents the degree (%) of splicing to the minor (B′) site, the abscissa represents the predicted stability of the stem in kilocalories, and each of the points represents the splicing of the noted mutated transcript.

efficient selection of the B′ acceptor; they act by stabilizing a stem-loop structure that encompasses the B acceptor site and its respective lariat-branch point A. Mutations that progressively stabilize this predicted stem result in a progressive increase in the use of the alternative B′ splice-acceptor (Figure 4). In the hGH-V transcript the loop is destabilized by the two base differences from the hGH-N transcript that happen to lie at the base of this stem. Thus, to efficiently use the B′ site, the lariat-branch point A that serves the B′ site and the two bases that sequester the B site in a stem-and-loop configuration must all be present. These results indicate that specific sequences dictate the relative use of the two competing splice-acceptor sites generating the functionally distinct 22 and 20 kDa hGH-N isoforms. Base substitutions at these positions explain the absence of alternative splicing at this site in the hGH-V transcript. Using this information we were able to design site-specific mutations into the hGH-N transcript in the region of the exon 3 splice-acceptor that force splicing toward the expression of only 22 or 20 kDa mRNAs.[9]

In contrast to the discrete nature of the proximal signal and its detailed characterization, a similar approach to the delineation of the distal signal failed to identify a unique region of importance. Instead, by expressing a series of chimeric hGH-N/hGH-V gene transcripts, we concluded that the distal splicing signal cannot be attributed to a single region. Instead, the "distal" signal appears to reflect the overall higher order structure of the hGH-N transcript.[10]

III. EXPRESSION OF hGH-RELATED GENES IN THE PLACENTA

A. hGH-V

The hGH-V gene was discovered when the entire hGH gene cluster was sequenced.[5] The existence of this gene was not appreciated before that time, and thus, no corresponding hormone was known. We initially tested the possibility that the hGH-V gene might encode a placental GH isohormone by transfecting the intact hGH-V gene into a fibroblast cell line. Analysis of this cell line demonstrated that the hGH-V gene could, in fact, be expressed by transfected cells and that it encoded the predicted 22 kDa hGH-V protein. This clearly demonstrated that hGH-V is not a psuedogene as speculated, but instead encodes a gestational GH.[11] We subsequently isolated a series of cDNA clones encoding hGH-V mRNA from a human placental library.[11] Sequence analysis revealed that several of these clones retain intron 4. Of note, the native reading frame of the hGH-V mRNA retaining intron 4, which we named hGH-V2, continues unobstructed throughout the retained intron 4 and well into exon 5. This alternatively spliced hGH-V2 mRNA predicted expression of a 26-kDa protein containing a 104 carboxy terminal amino acid sequence distinct from hGH-V.

The expression of the hGH-V gene was sublocalized within the placenta by Northern analysis and *in situ* histohybridization. Northern analysis of mRNA isolated from the four placental layers, amnion, chorion, decidua, and villi, demonstrated specific expression limited to the villi.[11] Expression of hGH-V was sublocalized within villi to the syncytiotrophoblastic epithelium by histohybridization using a complementary DNA (cDNA) probe specific for intron 4 of the hGH-V gene.[12]

We studied the protein expression of the hGH-V gene by analyzing the conditioned media of stably transfected cell lines. When the conditioned media of these hGH-V transfected cells were analyzed by Western blots or immunoprecipitation a major 22 kDa band and two minor bands of slightly slower migration were seen. These slower bands were shown to be N-linked glycosylation products of the 22 kDa form.[13] Of note, the 26 kDa hGH-V2 predicted from the mRNA analysis could not be detected. To further investigate whether this predicted protein could be expressed, we studied the expression of the hGH-V2 mRNA in *Xenopus* oocytes. In preliminary studies *Xenopus* oocytes were microinjected with mRNA transcripts encoding hGH-N, hGH-V, and hGH-V2. After labeling with ^{35}S-Met in both steady-state and pulse-chase experiments, hGH-N and hGH-V were synthesized and secreted into the culture media. In contrast, 26 kDa hGH-V2 was not present in the media, but instead remained associated with the cell lysates even after 6 h of chase with unlabeled methionine. Based upon these results and the presence of a hydrophobic domain in the novel C terminus of the hGH-V2, we suggested that hGH-V2 may be associated with the cell membrane.[14]

To determine the biological activity of hGH-V, we studied its binding to heterologous lactogen and somatogen receptors. hGH-V binds to both categories of receptors as does hGH-N. However, the ratio of somatogen to lactogen binding activity of hGH-V is 7.5-fold higher than that for hGH-N, predicting that hGH-V would be more purely somatogenic in its bioactivity profile.[15] To confirm this prediction, we studied the effect of hGH-V in parallel with hGH-N on weight gain in the hypophysectomized rat, the standard somatogen bioassay. Rats injected with hGH-V grew in parallel with those injected with hGH-N and both groups gained significantly more weight than those injected with control media. In the Nb2 cell mitogen assay, which measures lactogen bioactivity, hGH-V was about 20-fold less active than hGH-N.[16] These results demonstrate that the bioactivity profile of hGH-V parallels its receptor binding profile. We also showed that hGH-V is indistinguishable from hGH-N in its ability to bind to intact fat cells, to produce an insulin-like response, to induce refractoriness to insulin-like stimulation, and to induce lipolysis in the presence of glucocorticoid.[17] Because all the preceding assays were carried out in heterologous systems, we also studied the binding of hGH-V to the high-affinity GH binding protein in human plasma to be certain that it could bind to the ectodomain of the native hGH receptor. hGH-V was found to be equipotent with hGH-N for binding to this protein.[18] From these results, it is clear that hGH-V is the most purely somatogenic form of hGH and has the capacity to play a distinct role in the physiology of human pregnancy.

B. hCS-L

As was the case with hGH-V, the hCS-L gene was first discovered when the hGH gene cluster was sequenced.[5] This locus was assumed to represent an hCS pseudogene because a base substitution destroyed the GT dinucleotide essential for exon 2 splice-donor activity; the first base of this intron is mutated from G to A.[5] To further define the expression of the hCS-L gene, we asked whether it was expressed as an mRNA and if so what forms of mRNA were generated, and whether such mRNA levels were developmentally regulated. To specifically detect hCS-L in the background of the high levels of similar hCS-A and hCS-B mRNAs, we carried out an hCS-L mRNA-specific reverse transcriptase/polymerase chain reaction (RT/PCR). Samples of villous mRNA were analyzed throughout gestation. We discovered that hCS-L mRNA is expressed as early as 8 weeks of gestation.[19] In response to the inactivation of the normal splice-donor site for intron 2 two cryptic splice-acceptor sites are activated. Over the course of gestation, the relative representation of the two hCS-L mRNAs display a stable developmental profile — 88% in the larger form, 12% in the smaller form. To determine the structure of these two alternatively spliced products, each RT/PCR fragment was sequenced. We found that both processed transcripts used the same splice-donor site located 20 bases into intron 2. This site is present but cryptic in the other GH-related transcripts. The predominant splicing pattern joins this activated donor site to a novel splice-acceptor, which we call the L-acceptor, located 73 bp within exon 3. This splice maintains an open reading frame throughout the coding region. In the minor splicing pathway the same activated donor site splices to a novel acceptor located only 4 bases 3′ of the normal major acceptor site used in the hGH-N transcript. We have called this site the L′ acceptor. The presence of this new acceptor site was surprising, as its sequence diverges considerably from the usual consensus. However, usage of this splice-acceptor site also results in an open reading frame that would be lost had the normal acceptor site been used. Further studies identified three additional splicing pathways for hCS-L.[20] These findings clearly demonstrate that the hCS-L gene is expressed at the level of transcription and may encode one or more novel gestational hormones via alternatively spliced mRNAs.

C. hCS

The hCS-A and hCS-B genes (Figure 1) are coexpressed at high levels during gestation, and encode identical proteins. As is the case with the hGH-V and hCS-L genes, transcripts of the two hCS genes can be detected as early as 8 weeks of gestation, and their levels rise in parallel, suggesting common overall transcriptional regulatory features. By tracing the expression of the hCS-A and hCS-B genes during gestation using an assay which can differentiate and quantitate their respective mRNAs, we found that a change occurs in their relative concentrations.[19] At 8 to 10 weeks of gestation the ratio of hCS-A to hCS-B RNA is 1.5; by term gestation this ratio has increased to 5.0. The basis for this regulation is unknown, the main structural difference between the two genes being their different locations within the multigene cluster. The physiologic reason for this change is also obscure, as the two genes encode an identical hCS protein product. Similar to hGH-V2, hCS-A was found to retain its fourth intron in a small percentage of transcripts. This hCS-A2 mRNA maintains an open reading frame throughout its intron 4 and into exon 5 in a manner highly similar to that observed for the hGH-V2 mRNA. It is not known whether hCS-A2 encodes a protein *in vivo*.[19]

IV. AN hGH RECEPTOR ISOFORM IN THE PLACENTA

Because actual and potential GH-related ligands are expressed in the human placenta, we sought a locally expressed placental form of the GH receptor (GHR) by screening a placental cDNA library for evidence of an hGHR mRNA. A cDNA

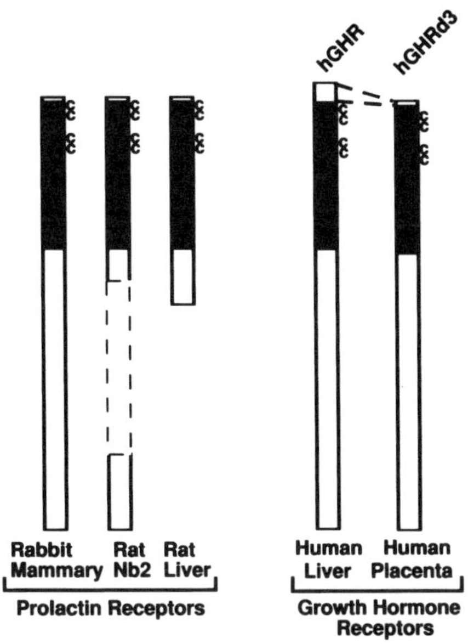

FIGURE 5. The structures of representative GH and Prl receptors. Three described isoforms of the Prl receptor differ solely by terminal or interstitial deletions in their cytoplasmic domains.[24] The interstitial deletion in the Prl receptor in the Nb2 lymphoma cell line is shown by the dotted region within the cytoplasmic domain. The two forms of the hGH receptor, hGHR and hGHRd3, are shown to the right. As detailed in the text, these two forms differ by alternative splicing of exon 3 of the hGHR gene transcript. The hGHRd3 transcript is the sole form detected in the placental villi. The cytoplasmic domain (below) is white, and the extracellular domain (above) is shaded and white. The region of the extracellular domain encoded by exon 3 is indicated by the dashed lines.

clone that encoded an hGHR transcript was found. However, sequence analysis revealed that this hGHR mRNA had an internal deletion; it was lacking 66 bases corresponding to nucleotides 72 through 137 of the liver receptor mRNA. This deletion represented a precise excision of sequences encoded by exon 3. The exon 3 deleted receptor isoform, which we called hGHRd3, is lacking 22 amino acids (Figure 5). This deletion begins six residues into the mature hGHR protein. By RT/PCR analysis, the hGHRd3 transcript was found to be the only form in the placental villi. Of interest, each of the other three layers of the placenta — chorion, amnion, and decidua — had a distinct distribution of the hGHR and hGHRd3 mRNA forms. A survey of human tissue culture cell lines revealed that the JEG-3 choriocarcinoma cell line had only the hGHR form, the hepatocellular carcinoma cell line Hep3B had only the hGHRd3 form, and both receptor isoforms were present in the IM9 lymphocyte cell line.[21]

To determine if hGHRd3 is a biologically active receptor, hGHR mRNA or hGHRd3 mRNA was microinjected into *Xenopus* oocytes. Receptors were detected via radioligand binding to the oocyte surface after overnight incubation.

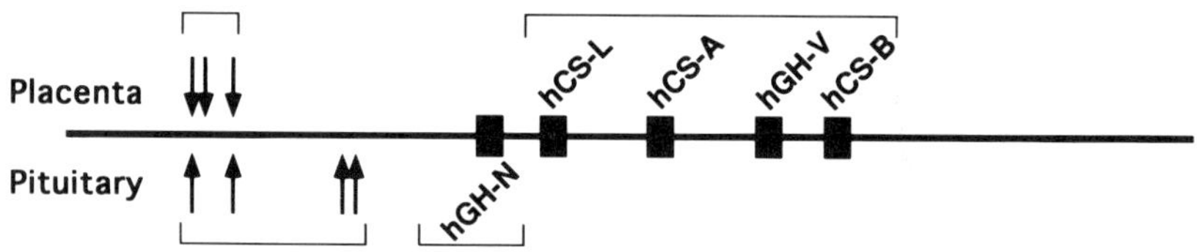

FIGURE 6. The DNaseI hypersensitive sites located 5′ to the hGH gene cluster. The positions of the tissue-specific DNaseI hypersensitive sites detected in nuclear chromatin isolated from pituitary (bottom) or placental villi (top) are shown by the vertical arrows.

Groups of such oocytes were studied for receptor binding affinity by competitive binding analysis between ^{125}I-hGH and a series of unlabeled potential ligands: hGH-N, 20 kDa hGH-N, hGH-V, hCS, and ovine Prl. In each case the competitive binding profiles were identical between hGHR and hGHRd3. Furthermore, rates of internalization of the two receptor isoforms were also identical. We concluded that hGHRd3 may be a functional receptor isoform.[22] The expression of potential ligands as well as the receptor in the placenta suggest that an autocrine/paracrine loop exists *in vivo* to mediate the bioactivity of the GH-related hormones in a local fashion during gestation.

V. A LOCUS CONTROL REGION REGULATES hGH-N GENE EXPRESSION *IN VIVO*

Although 500 bp of the hGH-N proximal promoter directs hGH gene expression in cultured pituitary cells, it does not direct significant levels of hGH-N gene expression in transgenic mice.[23] This discrepancy suggests the existence of additional, transcriptional regulatory sequences remote from the structural gene. To locate these regulatory elements, we analyzed 150 kb of placental chromatin encompassing the 40 kb hGH gene cluster for DNase I hypersensitive sites. This analysis was carried out on pure populations of nuclei from the syncytiotrophoblastic epithelial layer of the placental villi. A group of three tissue-specific hypersensitive sites were located between 27 and 32 kb 5′ of the hGH-N gene (Figure 6). A similar but more limited analysis in nuclei from a human pituitary GH-secreting adenoma revealed conservation of two of these three sites and two additional sites between 14 and 21 kb 5′ of hGH-N. A survey of chromatin from other tissues was negative for hypersensitive sites at these positions. A cosmid fragment containing hGH-N and 40 kb of contiguous 5′ flanking DNA, which included the hypersensitive sites was used to produce transgenic mice. Analysis of pituitary RNA from F1 offspring of lines established in this manner revealed that five out of five transgenic lines produced hGH-N mRNA in their pituitaries at levels comparable to the endogenous mouse GH mRNA. A similar result was obtained when 22.5 kb of flanking region (including only the more proximal hypersensitive sites) were retained 5′ to the hGH-N gene; four out of four lines expressed hGH mRNA in the pituitary at levels comparable to the endogenous mGH (murine growth hormone) mRNA. However, when the extent of the region 5′ to the hGH-N gene was further limited to 0.5 kb, 5 kb, or even 7.5 kb (three lines), the expression of the hGH-N transgene was either undetectable or present at only trace levels. When the distal set of hypersensitive sites was placed directly adjacent to the hGH-N gene, pituitary expression was detected in three out of three lines, although at only 1% of endogenous expression. From this analysis, we determined that sequences between −8 kb and −22.5 kb are essential for the activation of hGH-N transcription in the pituitary *in vivo.* DNase I mapping of chromatin from −8 to −22.5 kb in the pituitaries of transgenic mice carrying the full 40 kb of 5′ flanking region confirmed reformation of the two hypersensitive sites in the transgenic mouse pituitary. Thus, we located two sets of potential regulatory elements for the GH cluster. The hypersensitive sites at −27.5 to −32 kb permit predictable but low levels of pituitary expression and may function as locus boundary elements, or be more important for the expression of the placental set of genes. The hypersensitive sites at −8 to −22.5 kb may function as pituitary-specific locus control, enhancer elements, or both. These hypotheses are being tested. We conclude that the establishment of transcriptional competence of the hGH-N gene during the development and differentiation of the pituitary requires the participation of one or more of these major upstream regulatory element(s).[23]

REFERENCES

1. **Cooke, N. E., and Liebhaber, S. A.,** Molecular biology of the growth hormone-prolactin gene system, in *Vitamins and Hormones,* Litwack, G., Ed., Academic Press, San Diego, CA, in press.
2. **Cooke, N. E., Jones, B. K., Urbanek, M., Misra-Press, A., Lee, A. K., Russell, J. E., MacLeod, J. N., and Liebhaber, S. A.,** Placental expression and function of the human growth hormone gene cluster, in *Trophoblast Cells: Pathways for Maternal-Embryonic Communications,* Soares, M. J., Handwerger, S., and Talamantes, F., Eds., Springer-Verlag, New York, chap. 15, pp. 222–239, 1993.
3. **Cooke, N. E.,** Prolactin: basic physiology, in *Endocrinology,* 3rd ed., DeGroot, L. J., Ed., W. B. Saunders, Philadelphia, 1993, chap. 23.
4. **Barsh, G. S., Seeburg, P. H., and Gelinas, R. E.,** The human growth hormone gene family: structure and evolution of the chromosomal locus, *Nucleic Acids Res.,* 11, 3939, 1983.
5. **Chen, E. Y., Liao, Y.-C., Smith, D. H., Barrera-Saldaña, H. A., Gelinas, R. E., and Seeburg, P. H.,** The human growth hormone locus: nucleotide sequence, biology, and evolution, *Genomics,* 3, 479, 1989.
6. **Lewis, U. J.,** A naturally occurring structural variant of human growth hormone, *J. Biol. Chem.,* 253, 2679, 1978.
7. **Cooke, N. E., Ray, J., Watson, M. A., Estes, P. A., Kuo, B. A., and Liebhaber, S. A.,** Human growth hormone gene and the highly homologous growth hormone variant gene display different splicing patterns, *J. Clin. Invest.,* 82, 270, 1988.
8. **Estes, P. A., Cooke, N. E., and Liebhaber, S. A.,** A difference in the splicing patterns of the closely related normal and variant human growth hormone gene transcripts is determined by a minimal sequence divergence between two potential splice-acceptor sites, *J. Biol. Chem.,* 265, 19863, 1990.

9. **Estes, P. A., Cooke, N. E., and Liebhaber, S. A.,** A native RNA secondary structure controls alternative splice-site selection and generates two human growth hormone isoforms, *J. Biol. Chem.*, 267, 14902, 1992.
10. **Estes, P. A., Urbanek, M., Ray, J., Liebhaber, S. A., and Cooke, N. E.,** Alternative splice-site selection in the human growth hormone gene transcript and synthesis of the 20 kD isoform: role of higher order transcript structure, *Acta Paediat.*, Suppl., 399, 42–47, 1994.
11. **Cooke, N. E., Ray, J., Emery, J. G., and Liebhaber, S. A.,** Two distinct species of human growth hormone-variant mRNA in the human placenta predict the expression of novel growth hormone proteins, *J. Biol. Chem.*, 263, 9001, 1988.
12. **Liebhaber, S. A., Urbanek, M., Ray, J., Tuan, R., and Cooke, N. E.,** Characterization and histologic localization of human growth hormone-variant gene expression in the placenta, *J. Clin. Invest.*, 83, 1985, 1989.
13. **Ray, J., Jones, B. K., Liebhaber, S. A., and Cooke, N. E.,** Glycosylated human growth hormone variant, *Endocrinology*, 125, 566, 1989.
14. **Lee, A. K., MacLeod, J. N., Ray, J., Cooke, N. E., and Liebhaber, S. A.,** The human growth hormone-variant gene encodes a novel membrane-associated protein product, *Clin. Res.*, 296a, 1990.
15. **Ray, J., Okamura, H., Kelly, P. A., Cooke, N. E., and Liebhaber, S. A.,** Growth hormone-variant demonstrates a receptor binding profile distinct from that of normal pituitary growth hormone, *J. Biol. Chem.*, 265, 7939, 1990.
16. **MacLeod, J. N., Worsley, I., Ray, J., Friesen, H. G., Liebhaber, S. A., and Cooke, N. E.,** Human growth hormone-variant is a biologically active somatogen and lactogen with a bioactivity profile distinct from pituitary growth hormone, *Endocrinology*, 128, 1298, 1991.
17. **Goodman, H. M., Tai, L.-R., Ray, J., Cooke, N. E., and Liebhaber, S. A.,** Human growth hormone-variant produces insulin-like and lipolytic responses in rat adipose tissue, *Endocrinology*, 129, 1779, 1991.
18. **Baumann, G., Davila, N., Shaw, M. A., Ray, J., Liebhaber, S. A., and Cooke, N. E.,** Binding of human growth hormone (GH)-variant (placental GH) to GH-binding proteins in human plasma, *J. Clin. Endocrinol. Metab.*, 73, 1175, 1991.
19. **MacLeod, J. N., Lee, A. K., Liebhaber, S. A., and Cooke, N. E.,** Developmental control and alternative splicing of the placentally expressed transcripts from the human growth hormone gene cluster, *J. Biol. Chem.*, 267, 14219, 1992.
20. **Misra-Press, A., Cooke, N. E., and Liebhaber, S. A.,** Complex alternative splicing partially inactivates the human chorionic somatomammotropin-like (hCS-L) gene, *J. Biol. Chem.*, 269, 23220–23229, 1994.
21. **Urbanek, M., MacLeod, J. N., Cooke, N. E., and Liebhaber, S. A.,** Expression of a human growth hormone receptor (hGHR) isoform is predicted by tissue-specific alternative splicing of exon 3 of the hGH receptor gene transcript, *Mol. Endocrinol.*, 6, 279, 1992.
22. **Urbanek, M., Russell, J. E., Cooke, N. E., and Liebhaber, S. A.,** Functional characterization of the alternatively spliced, placental human growth hormone receptor, *J. Biol. Chem.*, 268, 19025, 1993.
23. **Jones, B. K., Monks, B. R., Liebhaber, S. A., and Cooke, N. E.,** Identification of a locus control region for the human growth hormone gene cluster, submitted.
24. **Kelly, P. A., Djiane, J., Postel-Vinay, M.-C., and Edery, M.,** The prolactin/growth hormone receptor family, *Endocr. Rev.*, 12, 235, 1991.

Chapter 2

GROWTH HORMONE-PROLACTIN RECEPTOR GENE FAMILY

Paul A. Kelly, Makoto Nagano, Athanassia Sotiropoulos, Jean-Jacques Lebrun, Phillipe Touraine, Laure Goujon, Helene Dinerstein, Fatima Ferrag, Helene Buteau, Alain Pezet, Nazario Esposito, Joëlle Finidori, Marie-Catherine Postel-Vinay, and Marc Edery

TABLE OF CONTENTS

I. INTRODUCTION

The anterior pituitary hormones growth hormone (GH) and prolactin (PRL), along with placental lactogens (PL) form a family of hormones which has been shown to be derived by duplication of an ancestral gene.[1] A wide spectrum of functions have been reported for prolactin (>85 in a various vertebrate species), whereas growth hormone is best known for its effects on the growth of skeletal and soft tissues and for its metabolic actions.

The initial step in the action of PRL and GH involves binding to a cell surface receptor. Although very little was known about the mechanism of action of PRL and GH, recent studies demonstrated that one of the initial events in the action of these hormones is the activation of a tyrosine kinase that in turn phosphorylates other proteins, including the receptor itself.

0-8493-8384-6/95/$0.00+$.50

The following sections describe the expanded family of GH/PRL/cytokine receptors, ligand binding determinants of the receptors, and the role of receptor dimerization and tyrosine phosphorylation in signal transduction.

II. IDENTIFICATION OF THE GROWTH HORMONE-PROLACTIN RECEPTOR GENE FAMILY

In 1987 William Wood's group at Genentech purified to homogeneity and sequenced the GH receptor (GHR) and binding protein (BP) in the rabbit and cloned the complementary DNAs (cDNAs) encoding the GH receptor in rabbits and humans.[2] Shortly thereafter, our group purified to homogeneity and sequenced the PRL receptor (PRLR) in rats and cloned its cDNA.[3] The first form of the PRLR to be identified was termed "short," because its cytoplasmic domain had only 57 amino acids, compared to the 350 amino acids of the GHR. Since that time, however, GHRs and PRLRs have been identified and characterized from a number of different species. As can be seen in Figure 1, short and long forms of both GHRs and PRLRs are now recognized. For GHRs, the short form is a BP that circulates in the blood and interacts with GH (i.e., GHBP); for PRLRs, the short form is membrane bound, representing a truncated version of the full length, long form of the receptor. Short and long forms of the PRLR are produced by alternative splicing of a single receptor gene in rat and mice.[4,5] Finally, an intermediate form of receptor, missing 198 amino acids in the cytoplasmic domain, is found in the rat lymphoma cell line Nb2.[6]

III. ORIGIN AND POTENTIAL ACTIONS OF GROWTH HORMONE-BINDING PROTEINS

Two separate mechanisms were proposed for the production of the GHBP. In mice and rats alternative splicing of a single primary transcript results in two distinct messenger RNAs (mRNAs). The 4.5 kb transcript encodes the full-length receptor and the 1.2 kb form encodes a truncated receptor, in which the transmembrane region has been replaced by a short hydrophilic sequence.[7,8] However, these murine species represent the only example of a separate transcript encoding the GHBP. The production of GHBP can also result from a second mechanism, such as specific proteolysis of the membrane receptor. In most species as only a single mRNA transcript (~4.5 kb) was identified by Northern blot analysis, the second mechanism is thought to explain the production of GHBP. In 1993 we demonstrated that Chinese hamster ovary (CHO) cells stably transfected with a cDNA encoding the long form of the rabbit GHR produced, in addition to the membrane-bound form of the receptor, high concentrations of soluble BP in the media.[9] When the cDNA encoding the GHR of the rat was stably expressed in CHO cells, only the membrane bound form was observed. These observations, coupled with the fact that N terminal sequence analysis of the GHBP was shown

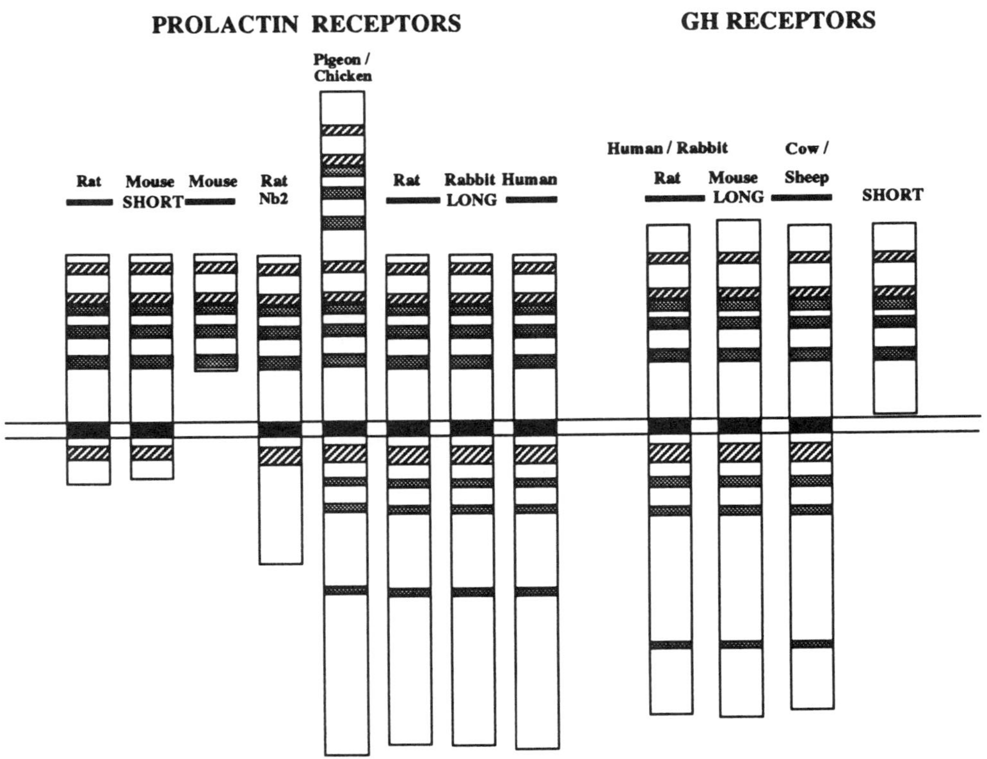

FIGURE 1. Schematic representation of various forms of PRLR and GHR. The short form of the PRLR from rat and mouse, the intermediate Nb2 form, and the long form of the PRLR in rat, rabbit, human, and birds (with the duplicated extracellular domain) are compared with the long and short (BP) forms of the GHR in human, rabbit, rat, mouse, and cow. The first and last amino acids of the mature protein are indicated. The transmembrane domain is represented by a black box. Regions of high (~70%) amino acid identity are cross-hatched, and those of moderate (40 to 60%) identity are stippled.

to be identical with the similar extracellular region of the GHR,[2] suggest that proteolytic cleavage is the main mechanism of production of BP in most non-murine species.

The biological functions of the GHBP remain to be clarified. This protein could act as a reservoir for GH in the circulation. Decreased degradation and metabolic clearance of GH was reported in a rat model when GH is bound to the BP.[10] Alternatively, the binding protein could serve to block GH actions, preventing further binding to membrane receptors.

IV. GROWTH HORMONE-BINDING PROTEIN IN HUMAN PLASMA

The GHBP in human plasma binds the hormone with a relative high affinity (5×10^8 M^{-1}) and a low capacity.[11,12] The affinity of the BP for the ligand is somewhat lower than that of the human GH receptor (hGHR).[13] The molecular weight of the GHBP is ~55,000, as evaluated by a number of different techniques. When complexed to the BP in adult plasma, GH remains immunoreactive. The proportion of GH bound to the BP was evaluated to be ~45%.

Growth hormone binding protein is measured by incubating serum plasma with ^{125}I-hGH, and different procedures for separating bound and free hormone are used, such as gel filtration,[11] high performance liquid chromatography,[12] and dextran-coated charcoal.[14] An "immunofunctional" assay and a radioimmuno-assay involving specific antisera to the GHBP were reported in the early 1990s.[15,16]

V. THE GROWTH HORMONE/PROLACTIN/CYTOKINE RECEPTOR FAMILY

The family that originally included GHR and PRLR has expanded to include receptors of a number of cytokines. Although the overall amino acid identity is low between members of this family, a significant (14 to 25%) identity exists over ~200 amino acids of the extracellular region of these receptors. In addition, two characteristic features are seen: the first is the presence of two pairs of cysteines, almost always found in the N terminal region of the molecule, which for each

---

FIGURE 2. Schematic representation of the GH/PRL/cytokine receptor family. Abbreviations: GHR: growth hormone receptor; PRLR: prolactin receptor; EPOR: erythropoietin receptor; IL-2R: interleukin-2 receptor; IL-3R: IL-3 receptor; IL-4R: IL-4 receptor; IL-5R: IL-5 receptor; GM-CSFR: granulocyte macrophage-colony stimulating factor receptor; IL-6R: IL-6 receptor; gp130: glycoprotein of M_r 130,000 (or β-subunit of IL-6R or oncostatin M receptor); IL-7R: IL-7 receptor; IL-9R: IL-9 receptor; MPL: myeloproliferative leukemia virus or orphan receptor of unknown ligand; CNTRF: ciliary neurotrophic factor receptor; LIFR: leukocyte inhibitory factor receptor; G-CSFR: granulocyte-colony stimulatory factor receptor. The plasma membrane is indicated by a stippled rectangle. The transmembrane region is shown in black. The thin black lines indicate the conserved cysteines and the thick black lines the WS × WS motif (tryptophan, serine, any amino acid, tryptophan, serine). Several receptors are formed by subunits, indicated 25 α, β, or γ.

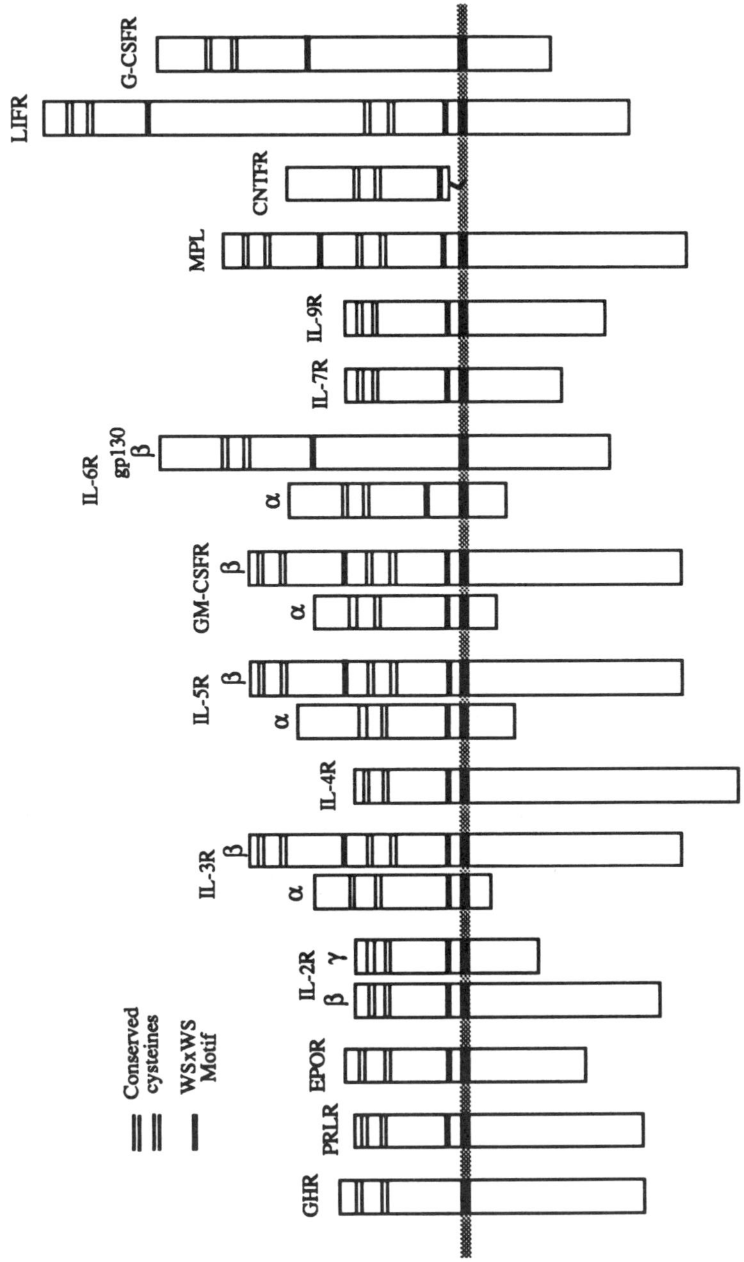
Conserved cysteines
WSxWS Motif
GHR
PRLR
EPOR
IL-2R
β
γ
IL-3R
α
β
IL-4R
IL-5R
α
β
GM-CSFR
α
β
IL-6R
α
gp130
β
IL-7R
IL-9R
MPL
CNTFR
LIFR
G-CSFR

pair were shown to be linked sequentially for the GHR. In addition, near the C terminal extremity of this homologous region is a highly conserved WS × WS motif (tryptophan, serine, any amino acid, tryptophan, serine) which is found in all members, except the GHR, in which some conservative substitutions occur. Finally, although very little conservation of primary sequence occurs in the cytoplasmic domains, three regions, known as Box 1, Box 2, and Box 3 are found in many members of this family. Figure 2 shows the expanded GH/PRL/cytokine receptor family. As can be seen, a number of receptors are formed by multiple subunits [interleukins (IL) 2, 3, 5, and 6, granulocyte macrophage-colony stimulating factor (GM-CSF), and probably leukocyte inhibitory factor (LIF) and ciliary neurotrophic factor receptor (CNTFR)].[17]

VI. EXPRESSION OF TWO FORMS OF PROLACTIN RECEPTOR TRANSCRIPT

Two approaches have been employed to measure the expression of the short and long forms of the PRLR. First, we developed a quantitative polymerase chain reaction (Q-PCR) in order to measure the absolute number of mRNA molecules encoding both forms of the PRLR. An essential aspect of Q-PCR is the construction of an internal control, as shown in Figure 3. Such a control RNA is reverse transcribed and amplified along with sample RNA under conditions that allow parallel responses. The details involved in the Q-PCR were described recently in a paper by Nagano and Kelly.[18] Using this technique, it was possible to detect as few as 500 molecules per microgram of total RNA. Sixteen tissues of adult female rats at two stages of the estrous cycle (proestrus and diestrus I) and the mammary gland of 20-d pregnant and 7-d lactating rats were examined. Receptor transcripts were detected in all tissues, with values ranging from $\sim 10^3$ molecules per microgram of total RNA in skeletal muscle to $\sim 10^7$ molecules per microgram of total RNA in liver, choroid plexus, and ovary. Most tissues expressed the long form transcript predominantly, although the thymus and kidney expressed both forms equally. The results indicate that the PRLR mRNA is ubiquitously but variably expressed in a tissue-specific manner and is clearly regulated by the hormonal environment associated with the stage of the estrous cycle, pregnancy or lactation.[18]

A second approach to measure the two receptor transcripts involves *in situ* hybridization. Probes specific to the intracellular domains of the short or long forms of the PRLR were prepared. The specificity of the signals was controlled by competition with excess unlabeled homologous probes or by hybridization with heterologous probes. Moreover, some tissues showed no expression of either form of receptor mRNA, and thus served as controls. The methods employed were recently described by Ouhtit et al.[19,20] *In situ* hybridization offers a means of directly identifying cells that express the mRNA of interest. Such an approach is especially well suited to the identification of short- and long-form transcripts in various tissues of the rat. The surprising finding, agreeing well with the results of

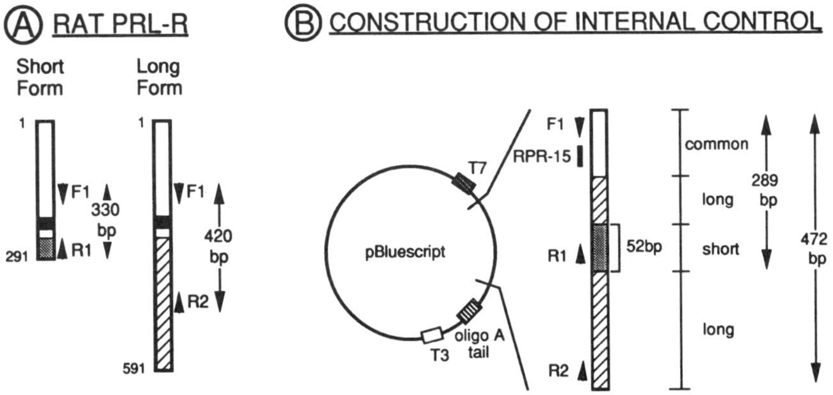

(C) PCR FRAGMENT SIZE (bp)

	Wild Type	Internal Control
Short	330	289
Long	420	472

FIGURE 3. Schematic presentation of the plasmid construction for the generation of cRNA. (A) Wild-type rat PRLR. The size of each form of the mature receptor is indicated on the left (amino acids), and the primer positions and the size of the PCR products are shown on the right (base pairs). Black box: transmembrane region; stippled box: specific region for short form; hatched box: specific region for long form. (B) Plasmid construction of internal control. pBluescript vector with T7 and T3 promoters and an oligo d(A) tail are shown at left. On the right, the structure of the insert is shown with the positions of the oligonucleotides used. Also indicated is the position of the insert structure with the size of amplicons of cRNA. (C) A comparison of the size (bp) of each amplicon.

Q-PCR, is that almost all tissues, and most frequently specific cells within a tissue, express varying levels of PRLR transcripts. Because mRNA expression was detected in many tissues for which PRL is not known to have an action, it is important to pursue future studies to correlate PRLR gene expression (by Q-PCR, *in situ* hybridization, and Western blot) with specific effects associated with each form of receptor.[19,20]

VII. BINDING DETERMINANTS OF GROWTH HORMONE, PROLACTIN, AND THEIR RECEPTORS

Homolog and alanine scanning mutagenesis of hGH originally identified a receptor-binding domain on the hGH molecule that involves two of the α-helices and the 54–74 loop region. Twelve residues were identified to form a patch on a two-dimensional structural model of hGH.[21] A similar approach was used to identify the binding determinants of the extracellular region of the GHR and the PRLR.[22,23] The binding domain of the receptors involves a region of ~100 amino acids, including the first four cysteines. This region was shown in structural models to form seven antiparallel β-strands grouped in a β-sheet sandwich.

VIII. THREE-DIMENSIONAL STRUCTURE, RECEPTOR DIMERIZATION, AND ACTION

Recent analysis of the three-dimensional crystal structure of the hGHBP and hGH confirmed that this complex forms a dimer with the ligand.[24] In a series of studies carried out by Kopchick's group[25] a second site of interaction between GH and its receptor was reported in the third α-helix of bovine GH (bGH). These authors mutated the glycine residue at position 119 into an arginine and established transgenic animals expressing the mutated gene. They expected to see giants, but to their great surprise, found dwarf mice. They reasoned that the first and fourth α-helices directly interacted with the receptor, but that residues in the third α-helix interacted with an unknown transmembrane protein that was necessary for functional activity. We now know, based on the elegant studies of De Vos et al.,[24,26] that the second protein is in fact a second receptor molecule, because the crystal structure studies, combined with the biochemical data from Wells' group, demonstrate that the extracellular binding protein exists as an unusual homodimer consisting of two molecules of receptor and one molecule of ligand. Thus, two receptor sites are found on hGH (identified as sites 1 and 2). Both sites bind to the same region of the hGHR. A sequential complex appears to form with the receptor first binding to site 1, after which a second receptor binds to site 2, followed by an interaction between the receptor molecules themselves that maintains the dimer complex.[26] Wells' Genentech group prepared a chimeric protein, consisting of the extracellular domain of the hGHR and the transmembrane and cytoplasmic domains of the G-CSF receptor. Stable transfection of this chimeric cDNA into FDC-P1 cells resulted in a biological test capable of measuring GH activity. GH mutants, with a reduced affinity for site 1, had a greatly reduced ability to stimulate proliferation of FDC-P1 cells containing the GH/G-CSF hybrid receptor compared to wild-type hGH. A mutant in site 2 of hGH (G120R, corresponding to glycine 119 of bGH), which was fully capable of binding the receptor, failed to activate proliferation, confirming the sequential two-site model of GH action.[27]

The evidence is strong that activation of PRLRs follows a similar mechanism involving dimerization. Monoclonal antibodies to the rat PRL receptor which are able to form receptor dimers were shown to be partial agonists.[28] More recently, using adjusted concentrations of second antibody to cross-link receptors, we were able to show that the same monoclonal antibodies are able to transduce full functional activity. In addition, monovalent Fab fragments which bind receptors are devoid of activity, but the addition of a second antibody restores the functional capacity.[29] In addition, the soluble, extracellular domain of the rat PRLR was shown to form 2:1 dimers with ovine PRL, identical to the complex formation between hGH and its binding protein.[22,30] Finally, G120R, which binds to the PRLR in the Nb2 lymphoma cell line, is able to block cell proliferation induced by PRL.[31] Final confirmation of the two-site model for PRL and dimerization of the receptor must await three-dimensional crystallography.

IX. FUNCTIONAL ACTIVITY OF GROWTH HORMONE AND PROLACTIN RECEPTORS

We developed an assay to measure the functional activity of transfected forms of the PRLR.[32] This assay involves the co-transfection of a PRL-responsive gene such as ovine β-lactoglobulin or rat β-casein coupled to a reporter gene, chloramphenicol acetyltransferase (CAT). Chinese hamster ovary cells are transiently transfected with the PRLR cDNA and the promoter/CAT fusion reporter gene. The transfected cells respond to PRL in the incubation media as measured by production of acetylated forms of chloramphenicol. We recently developed a similar assay to measure the functional activity of GHRs, using a fusion gene consisting of either ovine β-lactoglobulin/CAT or the serine protease inhibitor (SPI) 2.1/CAT.[33] Chinese hamster ovary cells transiently transfected with either of these constructs and the wild-type GHR respond to GH in the incubation media. The advantage of such assays is that they use transient transfection, and thus are well adapted to evaluate the cytoplasmic regions of the receptor required for the hormonal response.

X. THE CYTOPLASMIC DOMAIN AND BOX 1 IN SIGNAL TRANSDUCTION

Truncation and deletion mutants of GHRs were prepared and expressed in CHO cells. While the presence of ~50% of the cytoplasmic domain is sufficient for full activity of the PRLR, a similar mutant of the GHR is inactive.[34] A 25 amino acid region just inside the transmembrane domain is highly conserved between GHRs and PRLRs. Because of this we[3] originally proposed that this region may be important in the process of signal transduction. In fact, deletion of this juxtamembrane domain, either in GHR or PRLR leads to the complete loss of GH- or PRL-stimulated activity.[33,34] A more restrictive region within the juxtamembrane domain consisting of eight amino acids, known as Box 1, was identified in several members of the cytokine receptor family. As shown in Figure 4, deletion and alanine scanning mutagenesis confirmed that prolines of Box 1 are essential for the process of signal transduction for both receptors.[33,35] The hydrophobic amino acids (ILV) at the N terminal end of Box 1 of the GHR are also essential for functional activity.[33]

In another functional assay, utilizing the full-length GHR transfected into FDC-P1 cells, based on the original assay developed with the GH/G-CSF receptor

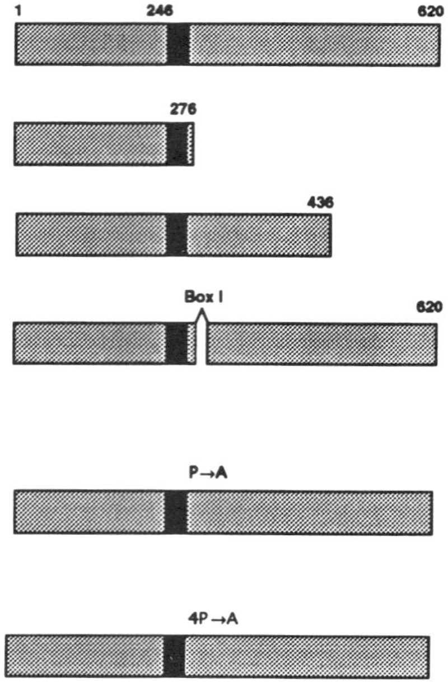

	Fold-Induction of		^{125}I hGH binding to cells	
	Spl/CAT Activity	ß-lacto/CAT Activity	K_d (nM)	sites/cell x 10^{-3}
Wild Type	3.7 ± 0.3	4.3 ± 0.3	0.29	143
T276	1.1 ± 0.1	1.3 ± 0.1	0.22	67
T436	1.3 ± 0.2	1.2 ± 0.1	0.32	238
Δ279-293	1.0 ± 0.1	0.9 ± 0.2	0.33	71
P282A	3.3 ± 0.4	ND	0.34	156
P283A	4.4 ± 0.8	ND	0.23	168
P285A	3.9 ± 0.4	ND	0.24	207
P287A	3.3 ± 0.3	ND	0.19	168
4P/A	1.4 ± 0.1	ND	0.23	115

hybrid, GH induced cell proliferation in the absence of IL-3.[36] Interestingly, these authors found that as little as 54 amino acids in the cytoplasmic domain were able to transmit a positive proliferative signal. Thus, only the juxtamembrane region containing Box 1 was necessary for the stimulation of cell proliferation, but it appears to be insufficient to stimulate GH-induced gene transcription.[33]

XI. GROWTH HORMONE- AND PROLACTIN-INDUCED TYROSINE PHOSPHORYLATION

Growth hormone has been shown to stimulate the phosphorylation of a protein with a molecular weight of ~120,000 in a number of different cell systems. Originally it was thought that the pp120 represented the GHR itself.[37] More recently studies clearly showed that an associated protein, and not the receptor, is the primary and initial tyrosine phosphorylated protein.[38] In addition, we demonstrated, using CHO cells stably expressing the rabbit GHR, that at least three tyrosine/phosphorylated proteins are induced following stimulation with GH. The receptor itself is also phosphorylated, but the degree of phosphorylation appears to depend on the cell system used. The functional role of the phosphorylated tyrosines in the various functional assays is currently being investigated.

Using Nb2 cells, our group and others[39,40] demonstrated the rapid stimulation of tyrosine kinase activity by PRL. We identified at least three tyrosine-phosphorylated proteins (pp120, pp97, and pp42) induced by lactogenic hormones. Phosphorylation of pp120 is maximal following incubation of cells with PRL for 1 min. Peak levels of pp97 and pp40 occur at somewhat later periods. The 42 to 44 kDa protein induced by both GH and PRL appears to be MAP kinase, a protein frequently involved in proliferation.

XII. JAK2, THE TYROSINE KINASE INVOLVED IN GROWTH HORMONE AND PROLACTIN ACTION

Although neither the GHR nor the PRLR contain a consensus sequence for adenosine/guanosine triphosphate binding or a kinase domain, a major advance in the field was made by the identification of JAK2 as a GHR-associated tyrosine

---

FIGURE 4. Functional activity of wild-type and mutant forms of the GHR. The wild-type receptor consists of 620 amino acids with a single transmembrane domain region (black). Truncated mutants (T276 and T436) have 6 and 166 amino acids in their cytoplasmic domain, respectively. The mutant Δ279–293 has 15 amino acids deleted in the cytoplasmic domain. The substitution mutants have one proline mutated to alanine in Box 1, the 4P mutant has four prolines (282, 283, 285, and 287), all changed to alanine. The functional activity (fold-induction of CAT activity) of each mutant is shown. Results represent the mean ± SEM of 5 to 14 independent experiments. All mutants were tested in the SPI/CAT functional test. The mutated forms T276, T436, and Δ279–293 were tested in the β-lactoglobulin/CAT test. Characteristics of the cell surface binding of [^{125}I] hGH to wild-type and mutant forms are shown. ND = not determined.

- **3 Family Members: JAK1, JAK2, & Tyk2**
- **2 Kinase domains**
- **50 % amino acid identity between family members**
- **Cytoplasmic localization**

FIGURE 5. Schematic representation of the JAK family of tyrosine kinase, including JAK1, JAK2, and Tyk2. The two kinase domains are shown cross-hatched and the other regions of homology are stippled.

kinase. JAK2 is a member of a family that also includes JAK1 and Tyk2 (Figure 5). All these proteins share the unusual feature of having two kinase domains. Complementary DNAs encoding these kinases were originally identified a few years ago,[41–43] although it was not known how they were activated. Stimulation of various cells expressing the GHR induced tyrosine phosphorylation of a protein with a relative molecular weight of 130,000, which could be immunoprecipitated with an antibody specific to JAK2.[44] Erythropoietin is also known to activate rapid tyrosine phosphorylation of a similarly sized protein, and this kinase was shown also to be JAK2.[45] In addition to the phosphorylation of JAK2, the GHR and erythropoietin receptors are also phosphorylated. Using GHR or erythropoietin receptor mutants, a membrane-proximal region of the cytoplasmic domain was shown to be important for biological activity, similar to the results presented earlier for GH on the role of the juxtamembrane region. In addition to GH and erythropoietin, JAK2 was shown recently to be the kinase that couples to the IL-3 receptor and the PRLR, and is probably implicated for GM-CSF, G-CSF, and interferon-γ (IFN-γ) receptors, as the first event in the process of signal transduction.[46,47] This is probably only the first step, however, in the signal transduction process that may involve other kinases, phospholipase C-γ, diacylglycerol, and several effector proteins (Figure 6).

XIII. DIRECT ACTIVATION OF A NUCLEAR TRANSCRIPTION FACTOR

A more direct path to the activation of transcription of specific genes may involve the phosphorylation of a cytoplasmic protein known as p91 or STAT (signal transducer and activator of transcription). Interferon-α/-β was shown to

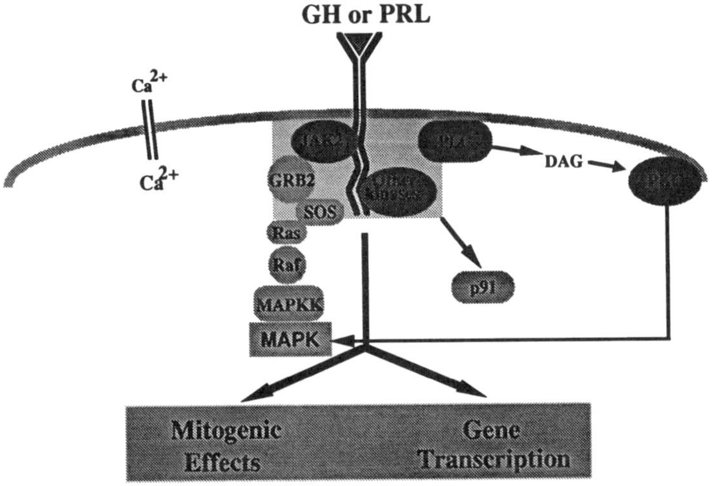

FIGURE 6. Schematic diagram representing potential pathways involved in GH and PRL signal transduction. Binding of the ligands activates the receptor-associated tyrosine kinase JAK2 and possibly other kinases. Cytoplasmic proteins containing SH_2 domains bind to the phosphorylated residues. GRB2 brings the guanine nucleotide exchange factor SOS to the receptor and results in sequential activation of Ras, Raf, and a series of protein kinases (mitogen-activated protein kinase, MAPK; MAP kinase kinase, MAPKK). A second signaling cascade involves activation of phospholipase C (PLC) and generation of diacylglycerol (DAG), which activates protein kinase C (PKC). A more direct pathway to gene activation was suggested in which the latent cytoplasmic protein p91 becomes tyrosine phosphorylated and then is translocated to the nucleus.

stimulate the transcription of specific target genes through a multimeric complex known as interferon-stimulated gene factor-3 (ISGF-3). This factor consists of cytoplasmic proteins p91/84 and p113, which are phosphorylated in response to IFN-α/-β by Tyk2 or JAK2, other members of the JAK family. Once activated, ISGF-3 combines with the cytoplasmic protein p48, and this complex migrates to the nucleus to activate transcription. Interferon-γ phosphorylates only one subunit (p91) of ISGF-3, probably via JAK1 or JAK2. An unexpected series of recent observations implicated p91/STAT, or more probably related family members, in the direct activation of growth factor responses. Thus, in addition to the now classical pathway involving activation of an intrinsic tyrosine kinase binding to SH2-containing proteins, activation of MAP kinase, and the transcription factor AP1, growth factors such as epidermal growth factor also phosphorylate, via their receptor kinases, a specific tyrosine (701) of p91/STAT, which when combined with an active *cis*-inducible factor, is able to activate transcription by direct interaction with response elements on target DNA.[48–52] It was recently demonstrated that GH is able to activate p91/STAT.[53] It will be most interesting to investigate how p91/STAT and other regulatory factors are involved in the transcriptional effects of GH and PRL.

REFERENCES

1. **Niall, H. D., Hogan, M. L., Sayer, R., Rosenblum, I. Y., and Greenwood, F. C.,** Sequences of pituitary and placental lactogenic and growth hormones. Evolution from a primordial peptide by gene duplication, *Proc. Natl. Acad. Sci. U.S.A.*, 68, 866, 1971.
2. **Leung, D. W., Spencer, S. A., Cachianes, G., Hammonds, R. G., Collins, C., Henzel, W. J., Barnard, R., Waters, M. J., and Wood, W. I.,** Growth hormone receptor and serum binding protein: purification, cloning and expression, *Nature,* 330, 537, 1987.
3. **Boutin, J. M., Jolicoeur, C., Okamura, H., Gagnon J., Edery, M., Shirota, M., Banville, D., Dusanter-Fourt, I., Djiane, J., and Kelly, P. A.,** Cloning and expression of the rat PRL receptor, a member of the GH/PRL receptor gene family, *Cell,* 53, 69, 1988.
4. **Shirota, M., Banville, D., Ali, S., Jolicoeur, C., Boutin, J. M., Edery, M., Djiane, J., and Kelly, P. A.,** Expression of two forms of the prolactin receptor rat ovary and liver, *Mol. Endocrinol.,*4, 1136, 1990.
5. **Davis, J. A. and Linzer, D. I. H.,** Expression of multiple forms of the prolactin receptor in mouse liver, *Mol. Endocrinol.,* 3, 674, 1989.
6. **Ali, S., Pellegrini, I., and Kelly, P. A.,** A prolactin dependent cell immune cell line (Nb2) expresses a mutant form of prolactin receptor, *J. Biol. Chem.,* 266, 20110, 1991.
7. **Smith, W. C., Kuniyoshi, J., and Talamantes, F.,** Mouse serum growth hormone (GH) binding protein has GH receptor extracellular and substituted transmembrane domains, *Mol. Endocrinol.,* 3, 984, 1989.
8. **Baumbach, W. R., Horner, D. L., and Logan, J. S.,** The growth hormone-binding protein in rat serum is an alternatively spliced form of the rat growth hormone receptor, *Genes & Dev.,* 3, 1199, 1989.
9. **Sotiropoulos, A., Goujon, L., Simonin, G., Kelly, P. A., Postel-Vinay, M. C., and Finidori, J.,** Evidence for generation of the growth hormone-binding protein through proteolysis of the growth hormone membrane receptor, *Endocrinology,* 132, 1863, 1993.
10. **Baumann, G., Amburn, K. D., and Buchanan, T. A.,** The effect of circulating growth hormone-binding protein on metabolic clearance, distribution, and degradation of human growth hormone, *J. Clin. Endocrinol. Metab.,* 64, 657, 1987.
11. **Baumann, G., Amburn, K., and Shaw, M.,** The circulating growth hormone-binding protein complex: a major constituent of plasma GH in man, *Endocrinology,* 122, 976, 1988.
12. **Tar, A., Hocquette, J. F., Souberbielle, J. C., Clot, J. P., Brauner, R., and Postel-Vinay, M. C.,** Evaluation of the growth hormone-binding proteins in human plasma using HPLC-gel filtration, *J. Clin. Endocrinol. Metab.,* 71, 1202, 1990.
13. **Hocquette, J. F., Postel-Vinay, M. C., Djiane, J., Tar, A., and Kelly, P. A.,** Human liver growth hormone receptor and plasma binding protein: characterization and partial purification, *Endocrinology,* 127, 1665, 1990.
14. **Amit, T., Barkey, R. J., Youdim, M. B. H., and Hochberg, Z.,** A new and convenient assay of growth hormone-binding protein activity in human serum, *J. Clin. Endocrinol. Metab.,* 71, 474, 1990.
15. **Carlsson, L. M. S., Rowland, A. M., Clark, R. G., Gesundheit, N., and Wong, W. L. T.,** Ligand-mediated immunofunctional assay for quantitation of growth hormone-binding protein in human blood, *J. Clin. Endocrinol. Metab.,* 73, 1216, 1991.
16. **Fairhall, K. M., Carmignac, D. F., and Robinson, I. C. A. F.,** Growth hormone (GH) binding protein and GH interactions in vivo in the guinea pig, *Endocrinology,* 131, 1963, 1992.
17. **Cosman, D.,** The hematopoietin receptor superfamily, *Cytokine,* 5, 95, 1993.
18. **Nagano, M. and Kelly, P. A.,** Tissue distribution and regulation of prolactin receptor gene expression: quantitative analysis by polymerase chain reaction, *J. Biol. Chem.,* 269, 13337, 1994.
19. **Ouhtit, A., Morel, G., and Kelly, P. A.,** Visualization of gene expression of short and long forms of prolactin receptor in the rat, *Endocrinology,* 133, 135, 1993.

20. **Ouhtit, A., Morel, G., and Kelly, P. A.,** Visualization of gene expression of short and long forms of prolactin receptor in rat reproductive tissues, *Biol. Reprod.,* 49, 528, 1993.
21. **Cunningham, B. C. and Wells, J. A.,** Rational design of receptor-specific variants of human growth hormone, *Proc. Natl. Acad. Sci. U.S.A.,* 88, 3407, 1991.
22. **Bass, S. H., Mulkerrin, M. G., and Wells, J. A.,** A systematic mutational analysis of hormone-binding determinants in the human growth hormone receptor, *Proc. Natl. Acad. Sci. U.S.A,* 88, 4498, 1991.
23. **Rozakis-Adcock, M. and Kelly, P. A.,** Identification of ligand binding determinants of the prolactin receptor, *J. Biol. Chem.,* 267, 7777, 1992.
24. **De Vos, A. M., Ultsch, M., and Kossiakoff, A. A.,** Human growth hormone and extracellular domain of its receptor: crystal structure of the complex, *Science,* 255, 257, 1992.
25. **Chen, W. Y., Wight, D. C., Mehta, B. V., Wagner, T. E., and Kopchick, J. J.,** Glycine 119 of bovine growth hormone is critical for growth-promoting activity, *Mol. Endocrinol.,* 5, 1845, 1991.
26. **Cunningham, B. C., Ultsch, M., De Vos, A. M., Mulkerrin, M. G., Clauser, K. R., and Wells, J. A.,** Dimerization of the extracellular domain of the human growth hormone receptor by a single hormone molecule, *Science,* 254, 821, 1991.
27. **Fuh, G., Cunningham, B. C., Fukunaga, R., Nagata, S., Goeddel, D. V., and Wells, J. A.,** Rational design of potent antagonists to the human growth hormone receptor, *Science,* 256, 1677, 1992.
28. **Elberg, G., Kelly, P. A, Djiane, J., Binder, L., and Gertler, A.,** Mitogenic and binding properties of monoclonal antibodies to the prolactin receptor in NB_2 rat lymphoma cells: selective enhancement by anti-mouse IgG, *J. Biol. Chem.,* 265, 14770, 1990.
29. **Rui, H., Lebrun, J. J., Kirken, R. A., Kelly, P. A., and Furrar, W. L.,** JAK2 activation and cell proliferation induced by antibody-mediated prolactin receptor dimerization, *Endocrinology,* in press.
30. **Hooper, K. P., Padmanabhan, R., and Ebner, K. E.,** Expression of the extracellular domain of the rat liver prolactin receptor and its interaction with prolactin, *J. Biol. Chem.,* 268, 22347, 1993.
31. **Fuh, G., Colosi, P., Wood, W. I., and Wells, J. A.,** Mechanism based design of prolactin receptor antagonists, *J. Biol. Chem.,* 268, 5376, 1993.
32. **Lesueur, L., Edery, M., Ali, S., Paly, J., Kelly, P. A., and Djiane, J.,** Comparison of long and short form of the prolactin receptor on prolactin-induced milk protein gene transcription, *Proc. Natl. Acad. Sci. U.S.A.,* 88, 824, 1991.
33. **Goujon, L., Allevato, G., Simonin, G., Paquereau, L., Le Cam, A., Clark, J., Nielsen, J. H., Djiane, J., Postel-Vinay, M. C., Edery, M., and Kelly, P. A.,** Cytoplasmic domains of the growth hormone receptor necessary for signal transduction, *Proc. Natl. Acad. Sci. U.S.A.,* 91, 957, 1994.
34. **Edery, M., Levi-Meyrueis, C., Paly, J., Kelly, P. A., and Djiane, J.,** A limited cytoplasmic region of the prolactin receptor proximal to the transmembrane domain is critical for signal transduction, *Mol. Cell Endocrinol.,* 102, 39, 1994.
35. **Pezet, A., Buteau, H., Edery, M., and Kelly, P. A.,** submitted for publication.
36. **Colosi, P., Wong, K., Leong, S. R., and Wood, W. I.,** Mutational analysis of the intracellular domain of the human growth hormone receptor, *J. Biol. Chem.,* 268, 12617, 1993.
37. **Carter-Su, C., Stubbart, J. R., Wang, X., Stred, S. E., Argetsinger, L. S., and Shafer, J. A.,** Phosphorylation of highly purified growth hormone receptors by a growth hormone receptor-associated tyrosine kinase, *J. Biol. Chem.,* 264, 18654, 1989.
38. **Wang, X., Möller, C., Norstedt, G., and Carter-Su, C.,** Growth hormone-promoted tyrosyl phosphorylation of a 121-kDa growth hormone receptor-associated protein, *J. Biol. Chem.,* 268, 3573, 1993.
39. **Rui, H., Djeu, J. Y., Evans, G. A., Kelly, P. A., and Farrar, W. L.,** Prolactin receptor triggering: evidence for rapid tyrosine kinase activation, *J. Biol. Chem.,* 267, 24076, 1992.

40. **Rillema, J. A., Campbell, G. S., Lawson, D. M., and Carter-Su, C.,** Evidence for a rapid stimulation of tyrosine kinase activity by prolactin in Nb2 rat lymphoma cells, *Endocrinology,* 131, 973, 1992.
41. **Firmbach-Kraft, I., Byers, M., Shows, T., Dalla-Favera, R., and Krolewski, J. J.,** Tyk2, prototype of a novel class of non-receptor tyrosine kinase genes, *Oncogene,* 5, 1329, 1990.
42. **Wilks, A. F., Harpur, A. G., Kurban, R. R., Ralph, S. J., Zurcher, G., and Ziemiecki, A.,** Two novel protein-tyrosine kinases, each with a second phosphotransferase-related catalytic domain, define a new class of protein kinase, *Mol. Cell. Biol.,* 11, 2057, 1991.
43. **Harpur, A. G., Andres, A. C., Ziemiecki, A., Aston, R. R., and Wilks, A. F.,** JAK2, a third member of the JAK family of protein tyrosine kinases, *Oncogene,* 7, 1347, 1992.
44. **Argetsinger, L. S., Campbell, G. S., Yang, X., Witthuhn, B. A., Silvennoinen, O., Ihle, J. N., and Carter-Su, C.,** Identification of JAK2 as a growth hormone receptor-associated tyrosine kinase, *Cell,* 74, 237, 1993.
45. **Witthuhn, B. A., Quelle, F. W., Silvennoinen, O., Yi, T., Tang, B., Miura, O., and Ihle, J. N.,** JAK2 associates with the erythropoietin receptor and is tyrosine phosphorylated and activated following stimulation with erythropoietin, *Cell,* 74, 227, 1993.
46. **Silvennoinen, O., Zitthuhn, B. A., Quelle, F. W., Cleveland, J. L., Yi, T., and Ihle, J. A.,** Structure of the murine Jak2 protein-tyrosine kinase and its role in interleukin 3 signal transduction, *Proc. Natl. Acad. Sci. U.S.A.,* 90, 8429, 1993.
47. **Lebrun, J. J., Ali, S., Sofer, L., Ulrich, A., and Kelly, P. A.,** Prolactin induced proliferation of Nb2 cells involves tyrosine phosphorylation of the prolactin receptor and its associated tyrosine kinase, *J. Biol. Chem.,* 269, 14021, 1994.
48. **Larner, A. C., David, M., Feldman, G. M., Igarashi, K. I., Hackett, R. H., Webb, D. S. A., Sweitzer, S. M., Petricoin, E. F., and Finbloc, D. S.,** Tyrosine phosphorylation of DNA binding protein by multiple cytokines, *Science,* 261, 1730, 1993.
49. **Ruff-Jamison, S., Chen, K., and Cohen, S.,** Induction by EGF and interferon-γ of tyrosine phosphorylated DNA binding proteins in mouse liver nuclei, *Science,* 261, 1733, 1993.
50. **Silvennoinen, O., Schindler, C., Schlessinger, J., and Levy, D. E.,** Ras-independent growth factor signaling by transcription factor tyrosine phosphorylation, *Science,* 261, 1736, 1993.
51. **Sadowski, H. B., Shuai, K., Darnell, J. E., Jr., and Gilman, M. Z.,** A common nuclear signal transduction pathway activated by growth factor and cytokine receptors, *Science,* 261, 1739, 1993.
52. **Shuai, K., Stark, G. R., Kerr, I. M., and Darnell, J. E., Jr.,** A single phosphotyrosine residue of stat 91 required for gene activation by interferon-γ, *Science,* 261, 1744, 1993.
53. **Meyer, D. J., Campbell, G. S., Cochran, B. H., Argetsinger, L. S., Larner, A. C., Finbloom, D. S., Carter-Su, C., and Schwartz, J.,** Growth hormone induces a DNA binding factor related to the interferon-stimulated 91-kDa transcription factor, *J. Biol. Chem.,* 269, 4701, 1994.

Chapter 3

RATIONAL DESIGN OF POTENT ANTAGONISTS TO THE HUMAN GROWTH HORMONE RECEPTOR

Abraham M. de Vos, Mark H. Ultsch, Anthony A. Kossiakoff, Germaine Fuh, Brian C. Cunningham, and James A. Wells

TABLE OF CONTENTS

I. INTRODUCTION

At the molecular level, the pharmacological and biological effects elicited by human growth hormone (hGH) result from ligand-induced activation of specific cellular receptors. Therefore, a promising approach toward modulation of these effects would be directed at intervention in the interaction of hGH with these receptors. Antagonists prevent receptor activation by interfering with the binding of the ligand, whereas agonists effect receptor activation even in the absence of hGH. Clearly, a prerequisite for any attempt at a rational design of antagonists or agonists is a thorough investigation of the target system. The first requirement is an understanding of the molecular mechanism whereby the receptor is activated, because the design strategy will depend on the details of this mechanism. Equally important is a detailed understanding of the ligand-receptor interactions at the atomic level in order to design molecules that can antagonize by blocking receptor binding or agonize by mimicking the natural ligand.

In the case of the hGH receptor mutational and structural studies have resulted in the elucidation of the activation mechanism, and this now represents perhaps the best-characterized system in terms of the extracellular events eliciting the cellular response.[1] This chapter reviews our recent work in these three areas and describes how the combination of the results in these fields made possible the design of highly effective GH receptor antagonists.

0-8493-8384-6/95/$0.00+$.50

II. MECHANISM OF ACTIVATION OF THE GROWTH HORMONE RECEPTOR

The cellular response to the presence of hGH in the extracellular environment is triggered by the activation of specific membrane-spanning GH receptors. Receptor activation is initiated by the ligand-binding event at the extracellular portion of the receptor (hGHBP). The detailed mechanism by which the hGH receptor transduces the extracellular stimulus through the cell membrane was recently elucidated.

The first indication of the composition of the hGH:hGHBP complex came from an analysis of crystals of the complex.[2] When these crystals were dissolved and the components separated by high-performance liquid chromatography, quantitation of the ratio of the components showed the composition to be one hormone molecule for every two molecules of receptor. Analysis of the complex in solution by means of gel filtration and titration calorimetry confirmed this composition, and demonstrated conclusively that a single hGH molecule binds two molecules of the extracellular domain of its receptor.[3] These results strongly suggest that the mechanism of activation of the GH receptor is ligand-induced dimerization. The observation that tightly binding analogues of hGH could only form 1:1 complexes and were biologically inactive supports this suggestion.[1] Further investigation of the complex in solution, using hGH variants that were defective in either of its receptor-binding sites, demonstrated that the two binding sites on the hormone have different affinities for the receptor. Formation of the complex proceeds in a sequential manner in which the high-affinity binding site on hGH (henceforth called site 1) is occupied first; only then can the second receptor bind to site 2 and complete the 1:2 complex (Figure 1). Furthermore, an excess of hGH will result in the dissociation of preformed 1:2 complexes, indicating that at least a subset of the hormone-binding determinants on the receptor is used in both site 1 and site 2. Consistent with the sequential mechanism of complex formation, a functional site 1 on the hormone is sufficient and necessary to induce dissociation of the 1:2 complex.[3]

Support for the biological relevance of the dimerization mechanism comes from the results of cell-based assays. In the absence of a good cellular signaling assay for hGH an assay was constructed based on the cytoplasmic signal transduction pathway of the granulocyte-colony stimulating factor (G-CSF) receptor. A myeloid leukemia cell line (FDC-P1), when transfected with the full-length G-CSF receptor, is stimulated to proliferate by G-CSF. In order to make these cells respond to hGH, a hybrid receptor was constructed in which the extracellular domain of the GH receptor was linked to a portion of the G-CSF receptor containing part of the extracellular domain, the transmembrane segment, and the cytoplasmic domain.[4] This hybrid receptor could be activated by hGH (Figure 2), as well as by monoclonal antibodies raised against the hGHBP, but not by their

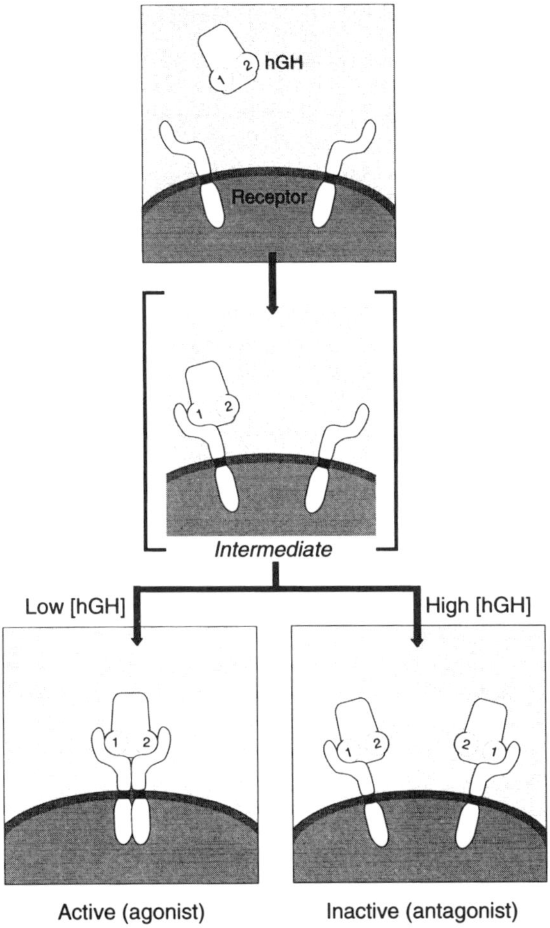

FIGURE 1. Mechanism of activation of the GH receptor. hGH first binds a receptor molecule with its site 1; addition of a second receptor to this intermediate complex produces the active hGH:(hGHBP)$_2$ complex. At high concentrations hGH binds all receptors through site 1 interactions, acting as an antagonist. (From *Science,* 256, 1677, 1992. © AAAS. With permission.)

monovalent Fab fragments. As expected from the solution studies with the hGHBP, antagonism of the proliferation activity can be achieved at high concentrations of hGH (Figure 2). Thus, the signal transduction pathway of the chimeric receptor requires receptor dimerization for activation, and cell-based assays with this receptor can be used to investigate the biological activity of hGH variants. Further evidence for the dimerization mechanism has been provided for IM9 lymphocytes and rat preadipocytes.[5,6]

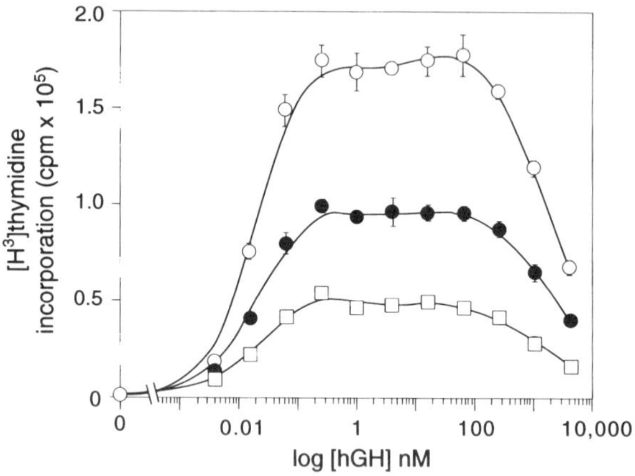

FIGURE 2. hGH-induced proliferation of FDC-P1 cells containing the hybrid hGH-mG-CSF receptor. Cells were added to 96-well plates at cell densities of 4×10^5 (open circles), 2×10^5 (closed circles), and 1×10^5 (closed squares) cells per millimeter, respectively. (From *Science*, 256, 1677, 1992. © AAAS. With permission.)

III. MAPPING OF HORMONE-RECEPTOR INTERACTIONS BY MUTAGENESIS

Substitution of an amino acid that is involved in important binding interactions by a functionally noncompatible residue will result in a decrease in the binding energy of complex formation. This observation forms the basis for the technique of alanine scanning mutagenesis, in which selected residues are replaced with alanine and the effect of these substitutions on binding affinity is measured. An important step in this process is the selection of the residues to be mutated, and the availability of structural information greatly simplifies the selection process. When the detailed three-dimensional structure of the complex is known, identification of the candidate residues can be based on visual inspection of ligand-receptor contacts. When only the structure of the individual components is available, the problem can still be reduced to a more manageable size by limiting the search to residues with exposed, solvent-accessible side chains. In the absence of any structural information whatsoever the selection can be guided only by more general criteria, such as the presence of charge or degree of hydrophilicity.

In the case of the high-affinity binding site on hGH, the initial decision as to which residues to mutate was based on the observation that bovine GH, human prolactin, and human placental lactogen do not bind to the hGH receptor. These homologues share between 30 and 85% sequence identity with hGH, and are therefore expected to have very similar three-dimensional structures. Later, a crude model of the porcine hormone (showing the structure to be a four-helix bundle) became available and helped in the interpretation of the results.[7] A set of 17 hGH variants was created by substituting segments of hGH with the corresponding regions of the homologues, leaving the buried hydrophobic residues unaltered.[8]

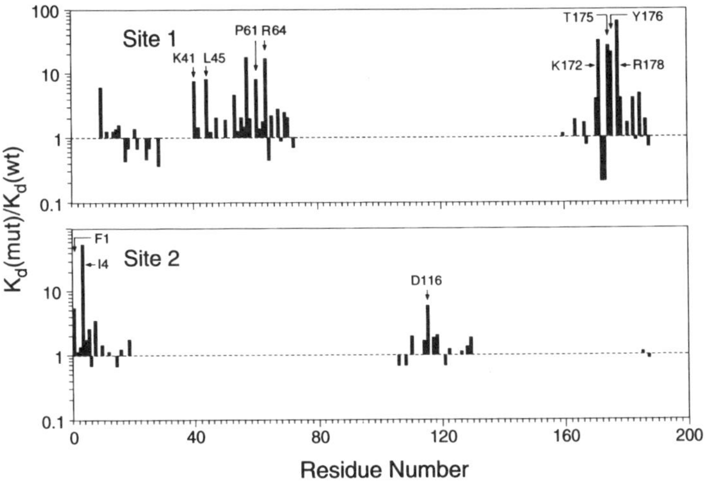

FIGURE 3. Change in binding relative to wild-type for alanine mutations in hGH. (Top) Residues in site 1; (bottom) residues in site 2.

Receptor binding of the resulting variants was assayed by competitive displacement of radiolabeled hGH and immunoprecipitation of the 1:1 hGH:hGHBP complex with Mab5. Six segment substitutions resulted in a significant decrease in hGH receptor binding, while not disrupting binding to a series of conformation-sensitive antibodies, and were therefore assumed to contain important receptor-binding determinants in wild-type hGH.[8] When mapped on the structure of the porcine hormone, these six segments were found to form a contiguous exposed patch on the surface. At this stage, the number of candidate binding determinants were reduced sufficiently to apply alanine scanning mutagenesis. The 62 residues in this patch whose side chains were larger than alanine were mutated one by one to alanine.[9] A group of 20 residues were identified for which replacement by alanine caused greater than twofold reduction in receptor binding; surprisingly, a few were found at which introduction of alanine *increased* binding. The resulting functional binding site was found to consist of the C terminal half of helix 4, some residues near the N terminus of helix 1, and the C terminal part of the first crossover connection between helices 1 and 2. Additional candidates for testing were identified from the crystal structure of the complex, mainly the first short helix in the connection between helices 1 and 2.[10] The set of residues, substitution of which by alanine reduced binding affinity more than sevenfold, are K41, L45, P61, R64, D171, K172, T175, F176, and R178.[11,12] Figure 3 (top) summarizes the results of the alanine scanning experiments for site 1.

For the analysis of the second receptor-binding site on hGH, a fluorescence assay was developed to follow receptor dimerization. A cysteine introduced into the hGHBP at position 237 was covalently modified with 5-iodoacetamidofluorescein.[3] Receptor dimerization brings the fluorescent groups on the hGHBP molecules in proximity to each other, because the site of attachment is near the beginning of the transmembrane segment of the intact receptor. Binding analysis demonstrated that this hGHBP variant has the same affinity for hGH as wild-type, and determination of the median effective concentration for hGH-induced dimerization gave a result comparable to the dissociation constant for the 1:2 complex as determined by immunoprecipitation (0.54 and 0.2 n*M*, respectively).[3] Furthermore, hGH analogues that only form 1:1 complexes do not induce homoquenching. This assay was used to evaluate a series of homologue- and alanine-scanning experiments directed toward identification of binding determinants in site 2, which was located in the N terminal region and the central part of helix 3. Greater than fivefold reduction in binding affinity resulted from alanine substitution at F1, I4, or D116 (Figure 3, bottom).[3]

The initial mutational analysis of the hGHBP was done when structural information was not yet available, and when hormone-induced dimerization had not yet been discovered. All charged residues were systematically changed to alanine; when a decrease in hormone binding occurred, the neighboring positions were probed as well.[13] Several important hormone-binding determinants were identified, notably W104 and P106. Recently, based on the structure of the complex, Clackson and Wells[14] undertook a comprehensive analysis of the residues in contact with the hormone in the first interface. This study confirmed the importance of the role of W104 and P106, and refined the earlier results for several of the other candidates. In addition, W169 was demonstrated to be as important for binding as W104.[14]

IV. CRYSTAL STRUCTURE OF THE hGH:(hGHBP)$_2$ COMPLEX

Only by a combination of structural information identifying the individual interactions in the ligand-receptor interface with mutagenesis data assessing the relative importance of these interactions can a detailed understanding at the atomic level of receptor activation be achieved. The X-ray crystallographic studies of the hGH:(hGHBP)$_2$ complex were greatly facilitated by the availability of large amounts of both hGH and the hGHBP. Both the hormone and the extracellular domain of the receptor are obtained in large quantities from efficient secretion systems that were developed in *Escherichia coli*.[15,16]

Crystallization of the hGH:(hGHBP)$_2$ complex was achieved after size-exclusion purification to remove any excess of the individual components.[2] The crystals were of good quality, diffracting to at least 2.7 Å resolution, and contained one 1:2 complex per asymmetric unit. The structure was determined at 2.8 Å resolution using the technique of multiple isomorphous replacement with two heavy atom derivatives by incorporating anomalous differences in the phasing procedure. The refined final structure has a crystallographic R factor of 0.194. Some segments were found to be disordered and are not part of the final structure;

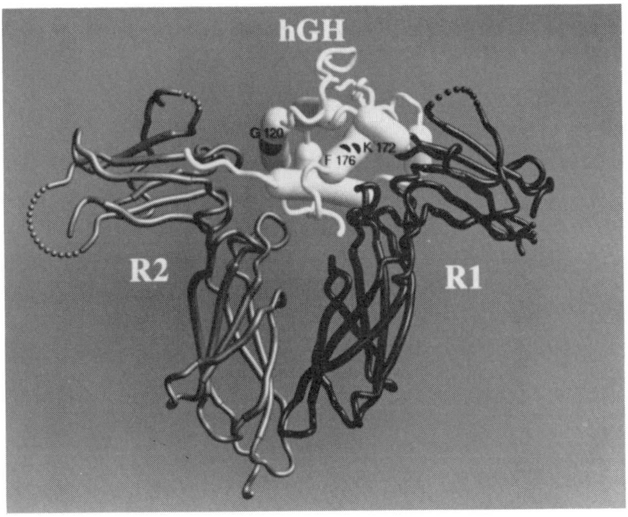

FIGURE 4. Structure of the hGH:$(hGHBP)_2$ complex. hGH is represented in white, the receptor bound to site 1 in black, and the receptor bound to site 2 in gray. The C termini of the receptors are near the bottom of the figure. (From *Science*, 256, 1677, 1992. © AAAS. With permission.)

these are residues 148–153 and 191 of hGH and residues 1–31, 56–61, 73–75, and 236–238 of each of the two receptor molecules.

The overall structure of the complex shows two hGHBPs bound at opposite sides of the hGH four-helix bundle, contacting each other near their C terminus (Figure 4).[10] The hGHBPs have a two-domain structure, each domain containing seven β-strands arranged similarly to immunoglobulin constant domains, with a single short linker segment in between. For each hGHBP, the contact surface with the hormone is formed by loops from each of the two domains. The surface of hGH in contact with the first receptor has a concave character and includes most of the exposed face of helix 4, part of helix 1, and residues in the first crossover connection. The second receptor-binding site is formed largely by the exposed side of helices 1 and 3, together with the amino terminus (Figure 5A). A considerable difference exists in the size of the contact surfaces: site 1 is about 1300 $Å^2$, whereas site 2 covers only about 850 $Å^2$. The third intermolecular interface between the stems of the two hGHBPs is about 500 $Å^2$. The difference in size of these contact surfaces supports the sequential mechanism of hGHBP binding; the smaller area in site 2 is apparently not sufficient to allow tight binding of the second hGHBP by itself, but when the first hGHBP is bound, the combined area of site 2 and the hGHBP interface is available to stabilize binding of the second hGHBP. The structure strongly supports dimerization as the mechanism of activation of the receptor: the fact that interface between the two hGHBPs is found in their C terminal stem, near where the cell membrane would be for the intact receptors, makes it almost inevitable that upon ligand binding the cytoplasmic

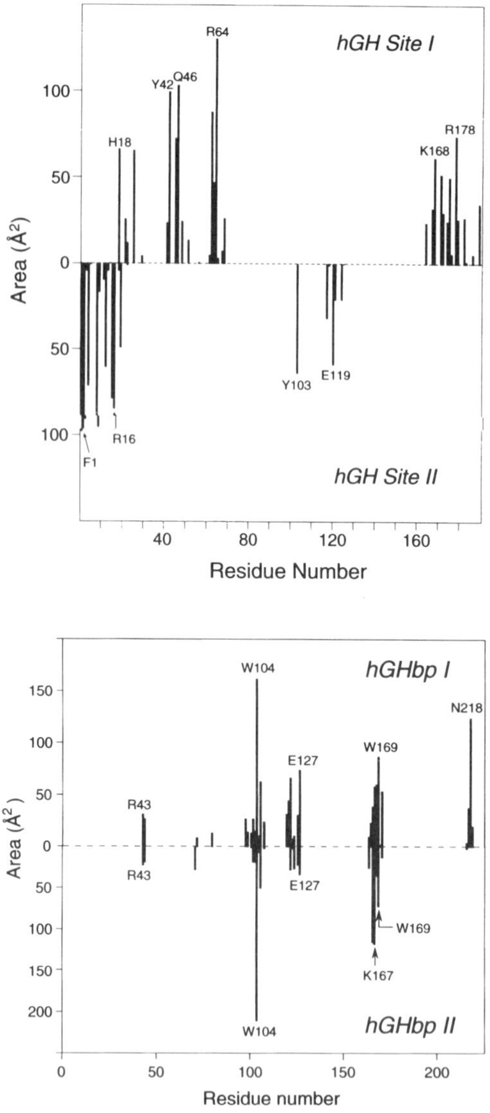

FIGURE 5. Residues making up the contact surface between hGH and its receptors. (A) Residues on the hormone that contribute to the contacts in size 1 (top) and site 2 (bottom). (B) Residues on the receptors contributing to the contact surface in site 1 (top) and site 2 (bottom). Note that the receptors use largely the same residues to interact with the hormone. (From De Vos, A. M. and Kossiakoff, A. A., *Curr. Opin. Struct. Biol.*, 2, 852, 1992. With permission.)

domains of the receptors would be in contact as well. Furthermore, it appears extremely unlikely that any conformational change in the extracellular part of the receptor could be transmitted to the cytoplasmic domain because the very C terminal dozen or so residues protrude from the compact C terminal domain, forming a flexible and partially disordered connection to the transmembrane segment (see Figure 4).

Analysis in greater detail of the interface with the hormone of each hGHBP reveals that each receptor uses the same set of residues for the interaction with hGH (Figure 5B).[10] For example, in each case R43, W104, P106, E127, and W169 are involved in contacts with the hormone. This finding is rather surprising, because hGH itself does not have any sequence duplication or symmetry. Apparently, the surface topography of each site on hGH allows the common set of receptor residues to find appropriate interactions; in order to achieve this, conformational differences between the two hGHBPs in some of the receptor loops that present the binding determinants are necessary, notably for W104 and W169. Calculation of the solvent accessible surface area each individual residue loses upon complex formation indicates that for both hGHBPs the most important binding determinants are W104 from the N terminal domain and W169 in the C terminal domain. In fact, these two tryptophans together account for about 20% of the total buried surface in site 1 and about 35% of that in site 2. In addition to these extensive hydrophobic interactions, a small number of intramolecular hydrogen bonds and salt bridges are found in the interfaces. In site 1 R43 of the receptor interacts with D171 and T175 of the hormone, and E127 with both hGH K41 and R167. In site 2 R43 of the second receptor hydrogen bonds to hGH N12, which also interacts with D126 of the hGHBP. hGH R16 and R19 interact with receptor E44 and Q166, respectively.

V. CORRELATION BETWEEN CRYSTAL STRUCTURE AND MUTAGENESIS DATA

Human growth hormone and its receptor presently represent one of the best-characterized systems available in terms of the details of ligand-receptor interactions at the atomic level. A wealth of mutational data describes the relative importance of the contacts observed in the three-dimensional structures. What conclusions can be drawn from a comparison of the structural and mutational data?

In general, the correspondence between the mutational data and the structure is very good. Of the 12 residues in hGH site 1 that show the largest decrease in binding affinity upon substitution by alanine, four are buried within the hormone and therefore presumably exert indirect effects (F10, F54, I58, and F176). Seven of the remaining eight (K41, L45, R64, Q68, D171, K172, T175, and R178) are involved in interactions with those residues that mutational analysis has identified as the most important hormone-binding determinants of the hGHBP (R43, W104, P106, E127, D164, W169). The only exception in this series is hGH R178, which the structure would not identify as an important binding determinant, although it is in van der Waals contact with hGHBP at G168 and V171. Alanine substitution of W104 or W169 of the hGHBP results in a dramatic loss of binding affinity, consistent with the large contact area seen in the structure.

On the other hand, the contact surface as defined by the three-dimensional structure covers a much larger area than suspected on the basis of the mutagenesis. Of about 30 important contact residues, 15 have virtually no effect on binding; indeed, about 80% of the total binding energy can be accounted for by only seven residues on the hormone.[12] Part of the reason might be that GH also binds to the prolactin (PRL) receptor; some of the residues that are not involved in GH receptor binding are important binding determinants for the PRL receptor.[11] The fact that the contact surface is much larger than the functional epitope is consistent with some theoretical and empirical studies on antigen-antibody complexes.[17,18]

VI. DESIGN OF POTENT ANTAGONISTS AGAINST THE hGH RECEPTOR

The rational design of antagonists against the GH receptor became possible when mechanistic, structural, and mutagenesis data were combined. An understanding of the sequential mechanism of receptor activation was required to find the approach upon which the design would be based. Knowledge of the structure of the complex made possible the identification of hormone-receptor interactions to interfere with, and consideration of the mutational data guided the actual choice and proved helpful in the optimization, of the antagonist.

Activation of the GH receptor is the result of receptor dimerization, in which each receptor in the activated complex uses the same residues for ligand binding. This suggests that antagonism could be achieved by saturating all free receptors as 1:1 complexes, thus preventing dimerization. Cell-based assays show that a large excess of hGH will antagonize the receptor by binding all available receptors with its site 1. We reasoned that a potent antagonist should simultaneously have a defective binding site for the second receptor and a high affinity for the first receptor. The design of an hGH mutant that was defective in its site 2 was based on the consideration that W104 of the hGHBP appeared to be the most important single residue for the interaction with the receptor. Inspection of the structure showed that the side chain of W104 of the second receptor fits snugly in a hole in the surface of hGH, and that one side of this hole is formed by the main chain of hGH G120. The close contact between the CA of this glycine and the side chain of W104 makes it improbable that a similar interaction can occur when a side chain is present at position 120 (Figure 6). Not surprisingly, an hGH mutant with R120 can no longer bind the second receptor, and is an efficient antagonist in cell-based assays[4] (Figure 7).

Next, binding affinity in site 1 was increased to optimize the antagonist. Based on the mutagenesis data, substitution of H21, R64, and E174 with alanine, lysine, and alanine, respectively, increases binding affinity for the first receptor 30 times.

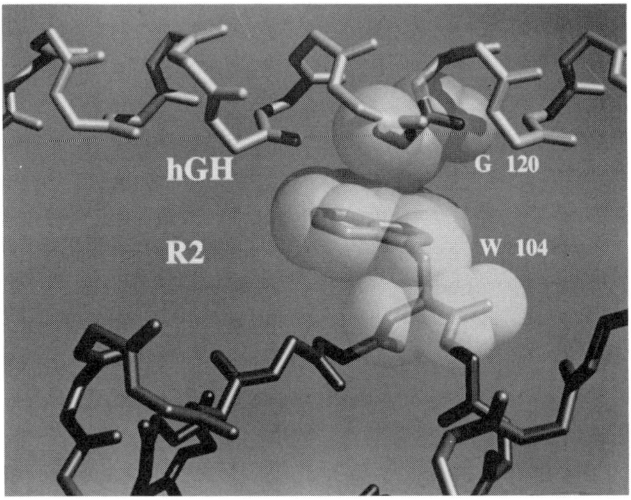

FIGURE 6. Close-up of the interaction of W104 of the receptor with site 2 on hGH. G120 on helix 3 of hGH (white) is in van der Waals contact with the side chain of W104 of the receptor (black). (From *Science,* 256, 1677, 1992. © AAAS. With permission.)

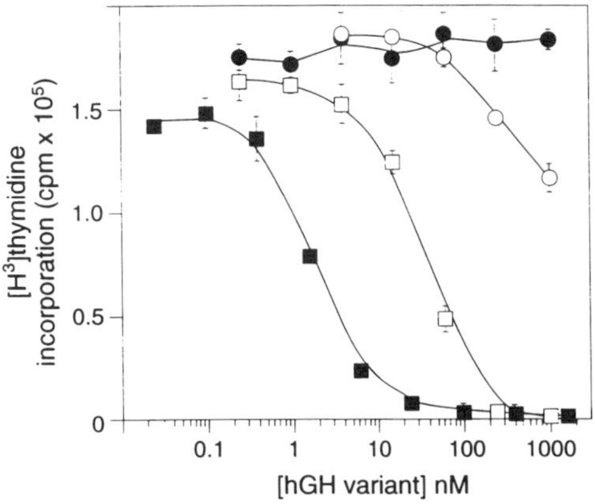

FIGURE 7. Antagonism of hGH-induced cell proliferation by hGH variants. Cells were incubated with 1 n*M* hGH and various concentrations of site 1 mutant K172A (filled circles), site 2 mutant G120R (open squares), combined mutant H21A/R64K/E174A/G120R (filled squares), or wild-type hGH (open circles). (From *Science,* 256, 1677, 1992. © AAAS. With permission.)

Incorporation of these three mutations in the R120 variant increased the efficiency of the antagonist by about tenfold (Figure 7).[4] Another potent antagonist was produced by a technique called phage display mutagenesis, in which comprehensive mutation of a selected set of residues is combined with selection for receptor binding. This variant contained a set of 15 mutations from wild-type, and bound about 400 times tighter to the first receptor.[19] Combination of the set of 15 mutations with the R120 mutation improved the initial antagonist tenfold.

VII. CONCLUSION

An understanding of the mechanism of hGH receptor activation at the molecular level, coupled with the structure of the activated complex and the mutational data guiding the interpretation of the observed ligand-receptor interactions, led to the successful design of potent antagonists. As yet, these antagonists are still variants of the wild-type ligand, and therefore relatively large protein molecules. The logical next steps along the way to a small-molecule antagonist would lie in a significant size reduction of the present molecule, followed by the design of organic mimics of the resulting peptide. This is obviously still a major hurdle to overcome, but the comparison of the structural and functional epitopes gives reason for cautious optimism because by far the greatest portion of the total binding energy was shown to reside in a relatively small number of interactions.

An alternative approach toward the design of small-molecule antagonists would be based on blocking the binding determinants on the ligand rather than the receptor. The structure of the complex shows the topography of the surface of hGH, and the mutagenesis data guide the selection of the interactions to interfere with. Combination of the structural and mutational data will again form the basis for the design of molecules that block important binding determinants on hGH.

The wealth of structural and functional data available for the hGH receptor system has opened up new possibilities for the rational development of small-molecule antagonists. It is too early to estimate how successful these attempts will be, but it is clear that exciting times lie ahead in the area of rational drug design.

REFERENCES

1. **Wells, J. A. and De Vos, A. M.,** Structure and function of human growth hormone: implications for the hematopoietins, *Annu. Rev. Biophys. Biomol. Struct.*, 22, 329, 1993.
2. **Ultsch, M., De Vos, A. M., and Kossiakoff, A. A.,** Crystals of the complex between human growth hormone and the extracellular domain of its receptor, *J. Mol. Biol.*, 222, 865, 1991.
3. **Cunningham, B. C., Ultsch, M., De Vos, A. M., Mulkerrin, M. G., Clauser, K. R., and Wells, J. A.,** Dimerization of the extracellular domain of the human growth hormone receptor by a single hormone molecule, *Science*, 254, 821, 1991.
4. **Fuh, G., Cunningham, B. C., Fukunaga, R., Nagata, S., Goeddel, D. V., and Wells, J. A.,** Rational design of potent agonists to the human growth hormone receptor, *Science*, 256, 1677, 1992.
5. **Silva, C. M., Weber, M. J., and Thorner, M. J.,** Stimulation of tyrosine phosphorylation in human cells by activation of the growth hormone receptor, *Endocrinology*, 132, 101, 1992.
6. **Ilondo, M. M., Damholt, A. B., Cunningham, B. C., Wells, J. A., Shymko, R. M., and De Meyts, P.,** *Endocrinology*, 134, 2397, 1994.

7. **Abdel-Meguid, S. S., Shieh, H.-S., Smith, W. W., Dayringer, H. E., Violand, B. N., and Bentle, L. A.,** Three-dimensional structure of a genetically engineered variant of porcine growth hormone, *Proc. Natl. Acad. Sci. U. S. A.,* 84, 6434, 1987.
8. **Cunningham, B. C., Jhurani, P., Ng, P., and Wells, J. A.,** Receptor and antibody epitopes in human growth hormone identified by homolog-scanning mutagenesis, *Science,* 243, 1330, 1989.
9. **Cunningham, B. C. and Wells, J. A.,** High-resolution epitope mapping of hGH-receptor interactions by alanine-scanning mutagenesis, *Science,* 244, 1081, 1989.
10. **De Vos, A. M., Ultsch, M., and Kossiakoff, A. A.,** Human growth hormone and extracellular domain of its receptor: structure of the complex, *Science,* 255, 306, 1992.
11. **Cunningham, B. C. and Wells, J. A.,** Rational design of receptor-specific variants of human growth hormone, *Proc. Natl. Acad. Sci. U.S.A.,* 88, 3407, 1991.
12. **Cunningham, B. C. and Wells, J. A.,** Comparison of a structural and a functional epitope, *J. Mol. Biol.,* 234, 554, 1993.
13. **Bass, S. H., Mulkerrin, M. M., and Wells, J. A.,** A systematic mutational analysis of hormone-binding determinants in the human growth hormone receptor, *Proc. Natl. Acad. Sci. U.S.A.,* 88, 4498, 1991.
14. **Clackson, T. and Wells, J. A.,** submitted.
15. **Chang, C. N., Rey, M., Bochner, B., Heyneker, H., and Gray, G.,** High level secretion of human growth hormone by Escherichia coli, *Gene,* 55, 189, 1987.
16. **Fuh, G., Mulkerrin, M. M., Bass, S., McFarland, N., Brochier, M., Bourrell, J. H., Light, D. R., and Wells, J. A.,** The human growth hormone receptor. Secretion from Escherichia coli and disulfide bonding pattern of the extracellular binding domain, *J. Biol. Chem.,* 265, 3111, 1990.
17. **Novotny, J., Bruccoleri, R. E., and Saul, F., A.,** On the attribution of binding energy in antigen-antibody complexes McPC 603, D1.3, and HyHEL-5, *Biochemistry,* 28, 4735, 1989.
18. **Davies, D. R., Sheriff, S., and Padlan, E. A.,** Antibody-antigen complexes, *Annu. Rev. Biochem.,* 59, 439, 1990.
19. **Lowman, H. B. and Wells, J. A.,** Affinity maturation of human growth hormone by monovalent phage display, *J. Mol. Biol.,* 234, 564, 1993.

Chapter 4

THE INSULIN-LIKE GROWTH FACTOR AXIS

Pinchas Cohen and Ron G. Rosenfeld

TABLE OF CONTENTS

I. ENDOCRINE REGULATION OF GROWTH

While multiple hormones influence growth, the primary regulator of postnatal somatic growth is growth hormone (GH). Growth hormone is secreted in a pulsatile manner from the anterior pituitary and is mainly under hypothalamic control, which in turn is regulated by neurotransmitters from higher cortical centers. The hypothalamic hormones, growth hormone releasing hormone and somatostatin, respectively stimulate and inhibit GH secretion.[1–3] In plasma the majority of GH is bound to a carrier protein termed GH binding protein (GHBP), which appears to be identical to the extracellular domain of the GH receptor.[4]

In the liver and other target tissues through interaction with its receptor, GH induces the production of somatomedins, or insulin-like growth factors (IGF-I and IGF-II).[5] These factors are found in plasma bound to a family of proteins called IGF binding proteins (IGFBPs).[6] Both IGFs and their binding proteins (primarily IGFBP-3) are reduced in GH deficiency.[1,2] The IGFs are thought to interact with target organs, such as growing cartilage, to induce growth; they also feedback on the pituitary to inhibit GH secretion.[7,8] It is still uncertain, however, whether all of the anabolic actions of GH are mediated via IGF-I. This cascade of growth control, known as the somatomedin hypothesis, is summarized in Figure 1.

0-8493-8384-6/95/$0.00+$.50

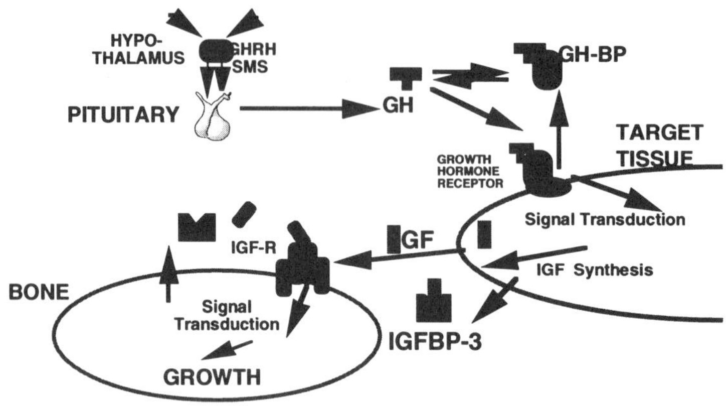

FIGURE 1. Growth regulation by GH and the IGFs. The complex cascade of growth regulation incudes growth hormone releasing hormone (GHRH), somatostatin (SMS), GH, GHBP, IGFs, insulin-like growth factor binding proteins (IGFBPs), and insulin-like growth factor receptors. (IGF-R).

II. THE INSULIN-LIKE GROWTH FACTOR AXIS

A. INSULIN-LIKE GROWTH FACTORS

Insulin-like growth factor -I and -II are two closely related peptide hormones of approximately 7 kD$^{\alpha}$ molecular weight. The IGFs were first identified in 1956 and were originally named sulfation factors or somatomedins.[5] The IGFs belong to a family of peptide hormones that include relaxin and insulin, and share a high degree of structural similarity with proinsulin. Like proinsulin they are composed of A, B, and C domains, but also include a "D" domain, which together form the mature IGF peptide. Both IGF-I and -II are synthesized with an additional extension peptide known as the "E peptide". In the liver and most other sites of IGF production this peptide is removed as part of the post-translational processing of IGF-I and -II. In some cases, however, larger forms of IGFs (particularly IGF-II) are secreted containing the E peptide. These larger molecular weight forms are also subject to glycosylation and may have apparent molecular weights of between 10 to over 20 kD$^{\alpha}$.[5,8]

Both IGF-I and -II have a complex gene structure, with the IGF-I gene spanning 95 kb (containing six exons) at the long arm of chromosome 12 and the IGF-II gene being made up of nine exons and having a total genomic size of 35 kb. Both genes are subject to multiple splicing and their messenger RNA (mRNA) species exist in several different sizes. The IGF-II gene is located on the short arm of chromosome 11 (near the insulin gene), in an area that appears to be paternally imprinted. This phenomenon may be related to the fact that IGF-II is a major fetal growth factor, although the postnatal roles of IGF-II are not well defined.[7,8]

The IGFs are important metabolic and mitogenic factors involved in cell growth and metabolism. Insulin-like growth factors are produced in the liver, in bone cells, and in other tissues, at least partially under GH control.[8] Circulating IGFs have direct (endocrine) effects on somatic growth and on the proliferation of many tissues and cell types, both *in vivo* and *in vitro*. However, the IGFs are also thought to be significant autocrine/paracrine factors involved in cellular proliferation.[5] Locally produced IGFs were demonstrated in bone, brain, prostate, muscle, mammary tissue, and other sites, where they are considered to be responsible for tissue growth and differentiation.[5,7,8]

B. INSULIN-LIKE GROWTH FACTOR RECEPTORS

The IGFs interact with specific receptors, designated as type I and type II IGF receptors (IGF-R), as well as with the insulin receptor.[9–11] The type I IGF-R binds IGF-I with high affinity, IGF-II with slightly lower affinity, and insulin with low affinity. The insulin receptor can bind IGF-I and -II, but with much lower affinity than insulin. The type II IGF-R receptor binds only IGF-II with high affinity, in addition to being a mannose-6-phosphate receptor.

The mitogenic effects of IGFs are thought to be mediated primarily through the type I IGF-R.[9] The metabolic effects of IGFs are probably mediated through the interactions of IGFs with the insulin receptor, but may involve interaction with the type I IGF-R, the insulin-IGF hybrid receptor, or both. It is unclear what function the type II IGF-R plays in mediating IGF action.

The type I IGF-R and the closely related insulin receptor are heterotetramers composed of a pair of α and β subunits which result from post-translational processing of a single gene product which encodes for the entire receptor. The two α subunits are linked by disulfide bonds and are primarily extracellular; they are involved in ligand binding. The β subunits are connected to the α subunit by disulfide bonds and function as intracellular tyrosine kinases. They undergo autophosphorylation after the interaction of the receptors with their ligands, followed by a conformational change.[9–11] Subsequently, these kinases appear to phosphorylate a cytoplasmic molecule known as the insulin-responsive substrate, which is involved in mediating many of the effects of insulin and IGFs.[11]

Evidence has recently emerged of the existence of a class of receptors for the IGF family, which has biological properties intermediate between the insulin and the type I IGF-R. Analysis of tissues in which these receptors appear to be common (such as the placenta), as well as transfection experiments, revealed that these receptors are composed of one insulin receptor α-β dimer and one type I IGF-R α-β dimer. This receptor was labeled the hybrid receptor, and appears to have high affinity for insulin as well as for IGFs.[9] The physiological role of the hybrid receptor remains elusive, but it may explain the potent insulin-like effects seen with intravenous administration of IGF-I to humans. While IGF-I binds to the insulin receptor with only 1 to 2% of the affinity of insulin, itself, it mediates hypoglycemia *in vivo* with 7 to 10% the effectivity of insulin.[5,11]

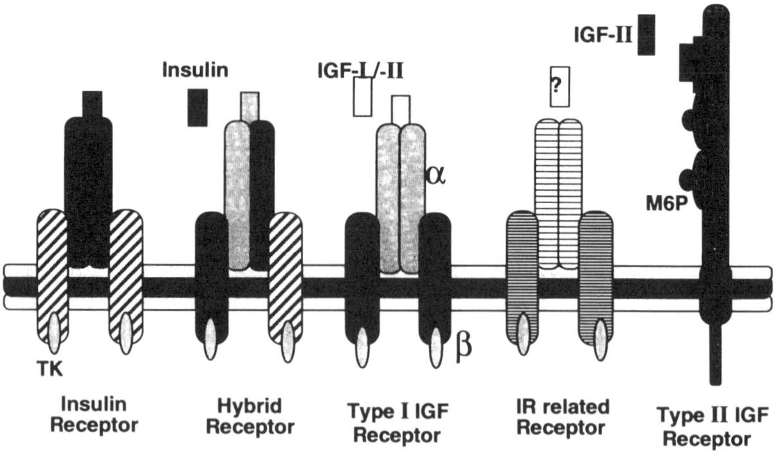

FIGURE 2. The IGF-Rs. IGF-I and -II bind to the α- subunit of their receptor. Signal transduction is mediated via the β-subunit, which includes a tyrosine kinase (TK).

The complementary DNA (cDNA) for a newly described member of this family of receptor was cloned recently, and due to its high homology with the insulin receptor was designated the insulin receptor-related receptor, or IRR. There are no known ligands for the IRR and both IGFs and insulin bind to it very poorly. It appears, however, to be expressed in a specific manner in renal and neural tissues and may have a role in fetal development. In chimeric transfection experiments it has been documented to be a very potent mediator of cellular proliferation.[10]

The type II IGF-R is structurally distinct, primarily binds IGF-II, but also serves as a receptor for mannose-6-phosphate-containing ligands.[9–11] It is not a member of the insulin receptor family, but rather displays homology to certain cytokine receptors. The receptor is 270 kDa in size and has 15 repeat extracellular domains. The type II IGF-R has a very short intracellular domain and and an unknown mechanism of signal transduction. It has been associated with changes in calcium influx and is reported to mediate cell motility. It was suggested that the type II IGF-R serves as a targeting mechanism to mediate lysosomal destruction of excess IGF-II during fetal life. Furthermore, it was shown in mice that this receptor is maternally imprinted and negatively controls the size of the fetus, further strengthening this hypothesis.[9,11] The IGFs and their cell surface receptors are depicted schematically in Figure 2.

C. INSULIN-LIKE GROWTH FACTOR BINDING PROTEINS

A recently recognized class of proteins with high affinities for the IGFs, the IGF binding proteins (IGFBPs), was shown to be involved in the modulation of the proliferative and mitogenic effects of IGFs on cells.[12,13] The molecular mechanisms involved in the interaction of the IGFBPs with the IGFs and their receptors remain unclear, but these molecules appear to regulate the availability of free IGFs for interaction with the IGF-R [13,14] The human IGFBP family consists of at least six proteins:[13–15]

IGFBP-1 is a 25-(kD2) protein found in high concentrations in amniotic fluid, and is also secreted by hepatocytes.[13]

IGFBP-2 has a molecular weight of 31 kDa, is found in serum cerebral spinal fluid, and seminal plasma, is secreted by many cell types, and is expressed in many fetal and adult tissues.[5]

IGFBP-3 is the major binding protein in postnatal serum and is synthesized by hepatocytes and other cells. In plasma, IGFBP-3 is found as part of a 150 kDa complex, which also includes an acid-labile subunit and an IGF molecule.[5,13]

IGFBP-4 is a 24-kDa protein that has been identified in serum and in seminal plasma, as well as in numerous cell types.[13,14]

IGFBP-5 is found in cerebral spinal fluid and in smaller amounts, in serum; it is also observed in rapidly growing fetal tissues.[14]

IGFBP-6 is found in cerebral spinal fluid and is produced by transformed fibroblasts; IGFBP-6 has relative specificity for IGF-II over IGF-I.[5,13–15]

A high degree of structural homology among the six cloned cDNAs for the binding proteins and remarkable sequence conservation across species was demonstrated. These binding proteins are tightly regulated by various endocrine factors and are uniquely expressed during ontogeny.[16] The IGFBPs are schematically depicted in Figure 3.

D. INSULIN-LIKE GROWTH FACTOR BINDING PROTEIN PROTEASES

Recently recognized as potential modulators of IGF action is a group of enzymes which are capable of cleaving IGFBPs.[17] First identified in pregnancy serum, IGFBP-3 proteolytic activity is responsible for the disappearance of intact IGFBP-3 from the serum of pregnant individuals, with no change in IGFBP-3 immunoreactivity.[18,19] These proteases were also reported in the serum of severely ill patients in states of cachexia, in patients with GH receptor deficiency, and in prostate cancer patients.[6,20] Seminal plasma contains an IGFBP-3 protease which has been identified as prostate-specific antigen. This IGFBP protease belongs to the kallikrein family.[21] The speculation was that the IGFBP proteases are important modulators of IGF bioavailability and bioactivity through their modification of the IGF carrier proteins. The proteolytic activity may play a role in regulating IGF availability at the tissue level by altering the affinity of the binding proteins for the growth factors, releasing free IGFs, and allowing increased receptor binding.[22] A theoretical model of such interactions is illustrated in Figure 4.

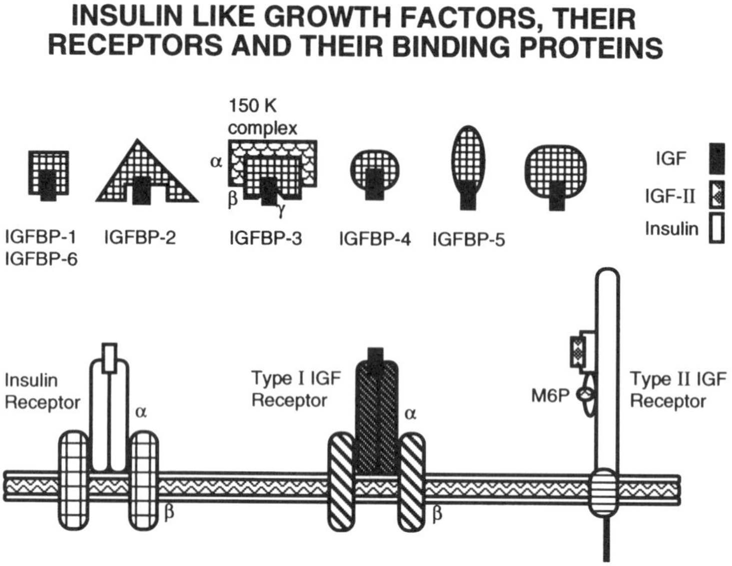

FIGURE 3. The IGF axis. IGF-I and -II bind to their receptors and to their binding proteins (IGFBPs). The 150 kDa complex includes the IGF (gamma), binding protein (beta), and the acid-labile (alpha) subunits.

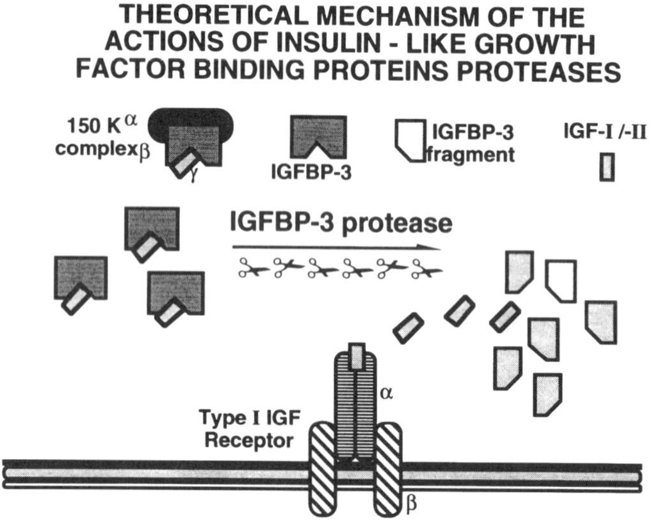

FIGURE 4. The actions of IGFBP proteases. IGFBPs are cleaved by a family to proteolytic enzymes (IGFBP proteases), which release bound IGFs to interact with their receptors.

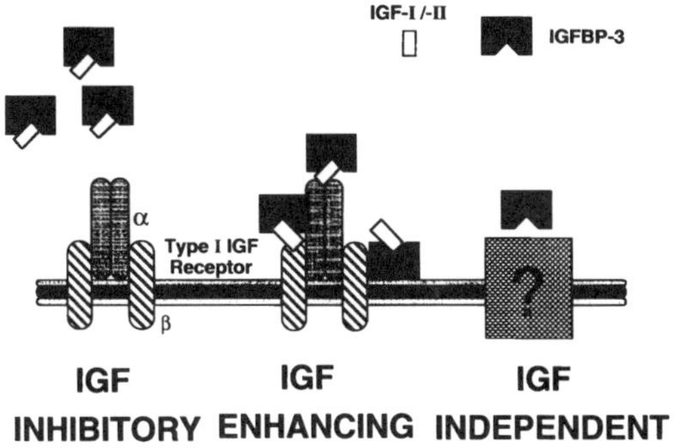

FIGURE 5. The actions of IGFBPs. IGFPBs can act in several ways to influence cell growth. Some actions may be IGF independent and their activity may involve cell-surface interactions. (Modified from Cohen, P. et al., *Growth Reg.*, 3, 23, 1993.)

E. ACTION OF INSULIN-LIKE GROWTH FACTOR BINDING PROTEINS

The molecular mechanisms involved in the interaction of the IGFBPs with the IGFs and their receptors remain enigmatic, but three possible models for their actions were suggested (see Figure 5). The first mechanism suggested for IGFBP action is that these molecules regulate the availability of free IGFs for interaction with the IGF-R.[13,14] Indeed, the addition of IGFBPs to many *in vitro* cell culture systems results in the inhibition of IGF actions within these experimental systems. Additionally, several trophic hormones were shown to suppress IGFBP production by their target cells. These include thyroid stimulating hormone inhibiting IGFBP-2 in thyroid cells, follicle stimulating hormone inhibiting IGFBP-3 in Sertoli cells, and follicle stimulating hormone inhibiting the production of an IGFBP produced by granulosa cells.[24–26] On the other hand, 1,25-vitamin D, which is inhibitory for bone cells, stimulates the production of IGFBP-4.[27] In these models it was assumed that the suppression or stimulation of an inhibitory IGFBP respectively stimulates or inhibits cell growth.[6]

In other systems, however, IGFBPs were demonstrated to enhance IGF action under circumstances that may involve cellular processing of the IGFBPs.[12–17,28] These IGF enhancing actions of IGFBPs were demonstrated for IGFBP-1, IGFBP-3, and IGFBP-5. It appears that a processing step, such as an affinity change involving proteolysis or phosphorylation, may be required for the activation of this function. Finally, as recently shown in several *in vitro* systems, including a cell transfection model, IGFBPs have an IGF-independent mechanism of cell inhibition.[29]

The mechanism by which IGFBPs independently interact with cells has not been fully elucidated. Both IGFBP-1 and IGFBP-2 contain an RGD sequence, which may allow them to interact with integrin receptors; IGFBP-3 does not contain such a sequence, but was recently shown to bind to specific binding sites on the cell membrane.[30]

The different actions of IGFBPs are illustrated in Figure 4.

III. THE INSULIN-LIKE GROWTH FACTOR AXIS IN DISEASE STATES

The serum levels of IGFs and IGFBPs are regulated ontogenically. The levels of IGF-I, -II and IGFBP-3 levels rise slowly throughout childhood and further increase during puberty.[2,6] Serum IGFs and IGFBP-3 levels remain stable during most of adult life and fall slowly during aging.[14] Levels of IGFBP-2 and IGFBP-4, on the other hand, appear to rise in aging people. The levels of IGFBP-1 are highest after birth and gradually decline afterwards.[6]

A. INSULIN-LIKE GROWTH FACTOR BINDING PROTEINS IN GROWTH DISORDERS

Serum concentrations of several IGF axis parameters are sensitive to GH secretory status.[6] Serum IGF-I, IGF-II, and IGFBP-3 levels are reduced in GH deficiency and are dramatically low in GH receptor deficiency.[6,14] Levels of IGF-I and IGFBP-3 are elevated in acromegaly.[6] Furthermore, serum IGF-I and IGFBP-3 rise in response to GH therapy in GH-deficient patients.[31] Nutrition plays a minor role in the regulation of serum IGFBP-3, although serum IGF levels are reduced in starvation and in poorly controlled diabetes.[32] In all of these situations the serum IGF-I and IGFBP-3 levels appear to positively correlate with GH and growth. However, IGF-II does not always maintain a direct relationship with these parameters, and may not be directly regulated by GH.

The role of IGF-I and particularly IGFBP-3 in the diagnosis of GH deficiency and other growth disorders is growing in its importance and is discussed elsewhere in this book.

While GH treatment consistently increases serum IGFBP-3, conflicting results were reported regarding the effects of IGF-I therapy. In animal models IGF-I appears to induce the production of IGFBP-3 and IGFBP-2.[34] However, in preliminary reports on the use of IGF-I in human patients with GHRD, IGF-I appears to stimulate growth without inducing the production of IGFBP-3.[33]

Serum IGFBP-2 levels appear to be inversely related to the GH secretory status. Concentrations of IGFBP-2 are increased in GH deficiency and GH receptor deficiency and reduced in acromegaly.[6] Furthermore, IGF-I treatment or elevated IGF-II levels (in certain tumors) are associated with suppression of GH secretion and increases in IGFBP-2.[6] Other IGFBPs do not seem to be directly related to GH status.

B. CHRONIC RENAL FAILURE

Reduced linear growth is common in children with chronic renal failure.[35] However, since serum GH levels are generally normal or elevated in such patients, it is likely that the growth failure is not directly GH related.[36] Patients with chronic renal failure also have normal serum IGF-I levels, indicating that both secreted GH and its receptor are functional.[37] The fact that these patients have elevated GH levels, despite their normal IGF levels, suggests a level of IGF resistance. Thus, the increased GH levels found in chronic renal failure may be due to a decrease in feedback inhibition of pituitary GH release by IGFs. This could happen in two ways: (1) a defect in the type 1 IGF-R, leading to reduced IGF action, or (2) an increase in serum IGFBPs, which could compete with the type 1 IGF-R for IGFs, and in this way inhibit IGF action. While no evidence for a defective type 1 IGF-R in chronic renal failure exists, recent data concerning IGFBP concentrations in these patients provide some interesting observations.[38,39]

Analysis of serum IGFBPs in patients with chronic renal failure revealed an increase in both low and high molecular weight IGFBPs.[38] Further characterization of these IGFBPs demonstrated that IGFBP-1 levels are elevated in chronic renal failure.[38,39] This is of interest because IGFBP-1 is generally found in amniotic fluid and maternal serum during gestation, and is not otherwise found in high amounts in serum.[40] A modest increase in serum IGFBP-3 concentrations in chronic renal failure was also reported.[41] It is attractive to hypothesize that both the elevated serum GH values and the growth failure characteristic of chronic renal failure are due to an inhibition of IGF action by an increase in IGFBP-1. It is also possible, however, that both the growth failure and the increased IGFBP-1 levels could be due to the poor nutritional status common in chronic renal failure as data indicate that IGFBP-1 regulation can be effected by nutritional status.[42] Furthermore, GH treatment in chronic renal failure results in increased growth velocity coupled with normalization of IGF and IGFBP serum levels.[41] It is clear that more studies are needed to determine if the growth failure associated with chronic renal failure can be linked to a disorder of IGFBP regulation, and to identify what the source of IGFBP-1 is in chronic renal failure.

C. THE INSULIN-LIKE GROWTH FACTOR AXIS IN DIABETES

In addition to their growth-promoting actions, IGFs have potent metabolic activities. Both IGF-I and -II were shown to be important regulators of glucose homeostasis *in vivo,* as well as *in vitro.*[43]

One of the postulated roles of IGFBPs is the prevention of the potential hypoglycemia that could arise from high plasma levels of free IGFs. It is thought that IGFBP-1 is the primary binding protein involved in modulating acute regulation of serum glucose levels by IGF peptides. Levels of serum IGFBP-1 measured by radiommunoassay are strongly correlated with metabolic state. *In vivo* in humans insulin appears to be the primary regulator of IGFBP-1; IGFBP-1 levels are inversely correlated to plasma insulin in essentially all conditions tested. Elevations in serum IGFBP-1 levels are seen in the hypoinsulinemia associated with fasting, type 1 diabetes mellitus, and exercise.[42,45,46] Reduced IGFBP-1 levels are seen in patients with insulinoma, after a meal or a glucose challenge, or during euglycemic-hyperinsulinemic clamps.[47] The apparent inverse correlation observed between serum IGFBP-1 and GH levels in GH deficiency and in acromegaly also may be related to the changes in insulin levels that are well described in these conditions. The relationship between serum IGFBP-1 and insulin is maintained with increasing age in relation to nutritional state and during the circadian rhythm.[42,48,49]

Uncontrolled insulin-dependent diabetes mellitus is often associated with growth retardation. Abnormalities of IGFs and IGFBPs are frequently reported in this condition. Total plasma IGF levels are reduced and serum IGFBP-3 levels were also reported to be low.[18,44] Both observations may represent a state of partial GH resistance in the poorly controlled diabetic patient. Additionally, elevated IGFBP-1 levels were suggested to have an inhibitory role on cartilage growth in insulin-dependent diabetes.[46] Acute administration of insulin to diabetic patients results in both metabolic normalization and in a fall of IGFBP-1 levels to normal.[45,47] Improved long-term control of diabetes is associated with normalization of both growth and IGFBP-1 levels.[45,47] Although a cause and effect relationship has not been firmly established, IGFBP-1 may function as a growth-inhibiting IGF antagonist in both uremic and diabetic sera.

D. THE INSULIN-LIKE GROWTH FACTOR AXIS IN PROLIFERATIVE DISORDERS

Several neoplastic conditions are characterized by altered expression of IGFs and related molecules.[48,49] Zapf et al.[50] detected IGF-I mRNA in tumors of neuroectodermal origin and in breast, bone, and liver tumors; IGF-II was reported to be expressed in excess in several benign and malignant conditions. High molecular weight IGF-II was associated with stromal tumors causing hypoglycemia.[51] Pekonen et al.[52] reported IGF-II mRNA in colon, breast, liver, and kidney tumors. Of note is that some cases of Wilms' tumor are associated with a duplication in the short arm of chromosome 11, which includes the site of the IGF-II gene. This gene is consequently overexpressed in the tumor and may stimulate tumor growth.[52]

The IGFBPs were recently suggested to be abnormally expressed in some tumors,[53,54] however, few of the reports demonstrating IGFBP expression in tumor cells used a comparable control tissue in a manner that convincingly demonstrated altered expression. Thus, although it is clear that most tumors secrete IGFBPs, no definitive examples of abnormalities in IGFBP expression are available.[6] Serum IGFBP-2 levels were reported to be increased in several malignancies, including leukemia and prostate cancer.[54] Whether this elevation in serum IGFBP-2 levels represents an early tumor marker, a paraneoplastic phenomenon or a tumor secretory product is not yet known.

The IGF axis is clearly involved in many cases of neoplastic transformation.[48] Reeve et al.[49] proposed that altered IGF-IGFBP-IGF-R balance in the autocrine/paracrine environment of the developing neoplasia may influence or promote tumor growth.

IV. SUMMARY

Over the last few years the medical and scientific literature has witnessed an explosion of information regarding the various components of the IGF axis. Insulin-like growth factors and related molecules are now believed to be critical elements in numerous physiological processes and key factors in several disease states. Undoubtedly, the coming years will bring even more new information on the physiology and pathology of these key cellular regulators. Furthermore, these discoveries are likely to lead to the increasing use of diagnostic tests involving IGFs, IGFBPs, and their proteases, as well as to therapeutic applications of these agents.

REFERENCES

1. **Frasier, F. D.,** Human pituitary growth hormone therapy in growth hormone deficiency, *Endocr. Rev.*, 4, 155, 1983.
2. **Hintz, R. L. and, Rosenfeld, R. G.,** Eds., *Growth Abnormalities,* Churchill Livingstone, New York, 1978.
3. **Cohen, P. and Rosenfeld, R. G.,** Growth problems in adolescence, in *Textbook of Adolescent Medicine,* MacAnarney, Kreipe, Orr, and Comerci, Eds., W. B. Saunders, Philadelphia, 1992, pp. 494–508.
4. **Baumann, G.,** Growth hormone-binding proteins, *Proc. Soc. Exp. Biol. Med.,* 202, 392, 1993.
5. **LeRoith, D., McGuinness, M., Shemer, J., Stannard, B., Lanau, F., Faria, T. N., Kato, H., Werner, H., Adamo, M., and Roberts, C. T.,** Insulin-like growth factors, *Biol. Signals,* 1, 173, 1992.
6. **Cohen, P., Fielder, P. J., Hasegawa, Y., Frisch, H., Giudice, L. C., and Rosenfeld, R. G.,** Clinical aspects of insulin-like growth factor binding proteins, *Acta Endocrinol. [Copenh.],* 124, 74, 1991.
7. **Nielsen, F. C.,** The molecular and cellular biology of insulin-like growth factor II, *Prog. Growth Factor. Res.,* 4, 257, 1992.
8. **Rotwein, P.,** Structure, evolution, expression and regulation of insulin-like growth factors I and II, *Growth Factors,* 5, 3, 1991.
9. **Oh, Y., Muller, H. L., Neely, E. K., Lamson, G., and Rosenfeld, R. G.,** New concepts in insulin-like growth factor receptor physiology, *Growth Reg.,* 3, 113, 1993.
10. **Zhang, B. and Roth, R. A.,** The insulin receptor-related receptor. Tissue expression, ligand binding specificity, and signaling capabilities, *J. Biol. Chem.,* 267, 18320, 1992.
11. **Nissley, P. and Lopaczynski, W.,** Insulin-like growth factor receptors, *Growth Factors,* 5, 29, 1991.
12. **Hintz, R. L. and Liu, F.,** Somatomedin plasma binding proteins, in *Growth Hormone and Other Biologically Active Peptides,* Pecile, A., and Muller, E. E., Eds., Excerpta Medica, Amsterdam, 1980, 133.
13. **Lamson, G., Giudice, L. C., and Rosenfeld, R. G.,** Insulin-like growth factor binding proteins: structural and molecular relationships, *Growth Factors,* 5, 19, 1991.
14. **Cohen, P., Ocrant, I., Fielder, P. J., Neely, E. K., Lamson, G., Oh, Y., Deal, L. C., Pham, H., Gargosky, S. E., Giudice, L. C., and Rosenfeld, R. G.,** Insulin-like growth factors: implications for aging, *Psychoneuroendocrinology,* 17, 335, 1992.
15. **Drop, S. L. S.,** Report on the nomenclature of the IGF binding proteins, *Endocrinology,* 130, 1736, 1992.

16. **Donovan, S. M., Oh, Y., Pham, H., and Rosenfeld, R. G.,** Ontogeny of serum insulin-like growth factor binding proteins in the rat, *Endocrinology,* 125, 2621, 1989.
17. **Lamson, G., Cohen, P., Guidice, L. C., Fielder, P. J., Oh, Y., Hintz, R. L., and Rosenfeld, R. G.,** Proteolysis of IGFBP-3 may be a common regulatory mechanism of IGF action in vivo, *Growth Reg.,* 3, 91, 1993.
18. **Davenport, M. L., Clemmons, D. R., Miles, M. V., Camacho-Hubner, C., D'Ercole, A. J., and Underwood, L. E.,** Regulation of serum insulin-like growth factor-I (IGF-I) and IGF binding proteins during rat pregnancy, *Endocrinology,* 127, 1278, 1990.
19. **Giudice, L. C., Farrell, E. M., Pham, H., Lamson, G., and Rosenfeld, R. G.,** Insulin-like growth factor binding proteins in maternal serum throughout gestation and in the puerperium: effects of a pregnancy-associated serum protease activity, *J. Clin. Endocrinol. Metab.,* 71, 806, 1990.
20. **Cotterill, A. M., Holly, J. M., Taylor, A. M., Davies, S. C., Coulson, V. J., Preece, M. A., Wass, J. A., and Savage, M. O.,** The insulin-like growth factor (IGF)-binding proteins and IGF bioactivity in Laron-type dwarfism, *J. Clin. Endocrinol. Metab.,* 74, 56, 1992.
21. **Cohen, P., Graves, H. C. B., Peehl, D. M., Kamarei, M, Giudice, L. C., and Rosenfeld, R. G.,** Prostrate specific antigen (PSA) is an IGF binding protein-3 (IGFBP-3) protease found in seminal plasma, *J. Clin. Endocrinol. Metab.,* 73, 491, 1991.
22. **Cohen, P., Peehl, D. M., Graves, H. C. B., and Rosenfeld, R. G.,** Biological effects of prostate specific antigen (PSA) as an IGF binding protein-3 (IGFBP-3) protease, *Proc. 74th Annu. Meet. Endocrine Society,* 1992, 960a.
23. **Cohen, P., Lamson, G., Okajima, T., and Rosenfeld, R. G.,** Transfection of the human IGFBP genes into Balb/c fibroblasts: a model for the cellular functions of IGFBPs, *Growth Reg.,* 3, 23, 1993.
24. **Bachrach, L. K., Liu, F. R., Burrow, G. N., and Eggo, M. C.,** Characterization of insulin-like growth factor-binding proteins from sheep thyroid cells, *Endocrinology,* 125, 2831, 1989.
25. **Smith, E. P., Dickson, B. A., and Chernausek, S. D.,** Insulin-like growth factor binding protein-3 secretion from cultured rat Sertoli cells: dual regulation by follicle stimulating hormone and insulin-like growth factor-I, *Endocrinology,* 127, 2744, 1990.
26. **Adashi, E. Y., Resnick, C. E., Hurwitz, A., Ricciarelli, E., Hernandez, E. R., and Rosenfeld, R. G.,** Ovarian granulosa cell-derived insulin-like growth factor binding proteins: modulatory role of follicle stimulating hormone, *Endocrinology,* 128, 754, 1991.
27. **Scharla, S. H., Strong, D. D., Mohan, S., Baylink, D. J., and Linkhart, T. A.,** 1,25-Dihydroxy vitamin D3 differentially regulates the production of insulin-like growth factors-I (IGF-I) and IGF-binding protein-4 in mouse osteoblasts, *Endocrinology,* 129, 3139, 1991.
28. **Clemmons, D. R.,** Insulin-like growth factor binding proteins: roles in regulating IGF physiology, *J. Dev. Physiol.,* 15, 105, 1991.
29. **Cohen, P., Lamson, G., Okajima, T., and Rosenfeld, R. G.,** Transfection of the human insulin-like growth factor binding protein-3 gene into Balb/c fibroblasts inhibits cellular growth, *Mol. Endocrinol.,* 7, 380, 1993.
30. **Oh, Y., Muller, H. L., Lamson, G., and Rosenfeld, R. G.,** Insulin-like growth factor (IGF)-independent action of IGF-binding protein-3 in Hs578T human breast cancer cells. Cell surface binding and growth inhibition, *J. Biol. Chem.,* 268, 14964, 1993.
31. **Baxter, R. C. and Martin, J. L.,** Radioimmunoassay of growth hormone-dependent insulin-like growth factor binding protein in human plasma, *J. Clin. Invest.,* 78, 1504, 1986.
32. **Batch, J. A. and Werther, G. A.,** Changes in growth hormone concentrations during puberty in adolescents with insulin dependent diabetes, *Clin. Endocrinol.,* 36, 411, 1992.
33. **Vaccarello, M. A., Diamond, F. B., Guevara-Aguirre, J., Rosenbloom, A. L., Fielder, P. J., Gargosky, S., Cohen, P., Wilson, K., and Rosenfeld, R. G.,** Hormonal and metabolic effects and pharmacokinetics of recombinant insulin-like growth factor-I in growth hormone receptor deficiency/Laron syndrome, *J. Clin. Endocrinol. Metab.,* 77, 273, 1993.

34. **Zapf, J., Hauri, C., Waldvogel, M., Futo, E., Hiasler, H., Biz, K., Guler, H. P., Schmid, C., Freosch, and E. R.,** Recombinant human IGF-I induces its own specific carrier proteins in hypophysectomized and diabetic rats, *Proc. Natl. Acad. Sci. U.S.A.*, 86, 8313, 1989.
35. **Betts, P. R. and Magrath, G.,** Growth patterns and dietary intake of children with chronic renal insufficiency, *Br. Med. J.*, 2, 189, 1974.
36. **Samaan, N. A. A. and Freeman, R. M.,** Growth hormone levels in severe renal failure, *Metabolism*, 19, 102, 1970.
37. **Powell, D. R., Rosenfeld, R. G., Baker, B., Liu, F., and Hintz, R. L.,** Serum somatomedin levels in adults with chronic renal failure: the importance of measuring insulin-like growth factor I (IGF-I) and IGF-II in acid chromatographed uremic serum, *J. Clin. Endocrinol. Metab.*, 63, 1186, 1986.
38. **Lui, F., Powell, D. R., and Hintz, R. L.,** Characterization of insulin-like growth factor-binding proteins in human serum from patients with chronic renal failure, *J. Clin. Endocrinol. Metab.*, 70, 620, 1990.
39. **Lee, P. D. K., Hintz, R. L., Sperry, J. B., Baxter, R. C., and Powell, D. R.,** IGF binding proteins in growth-retarded children with chronic renal failure, *Pediatr. Res.*, 26, 308, 1989.
40. **Hall, K., Lundin, G., and Povoa, G.,** Serum levels of the low molecular weight form of the insulin-like growth factor binding protein in healthy subjects and patients with growth hormone deficiency, acromegaly and anorexia, *Acta Endocrinol.*, 118, 321, 1988.
41. **Tonshoff, B., Mehis, O., Heinrich, U., Blum, W. F., Ranke, M. B., and Schauer, A.,** Growth-stimulating effects of recombinant human growth hormone in children with end-stage renal disease, *J. Pediatr.*, 116, 561, 1990.
42. **Busby, W. H., Snyder, D. K., and Clemmons, D. R.,** Radioimmunoassay of a 26,000-dalton plasma insulin-like growth factor binding protein: control by nutritional variables, *J. Clin. Endocrinol. Metab.*, 67, 1225, 1988.
43. **Rosenfeld, R. G.,** Somatomedin action and tissue growth-factor receptors, in *Acromegaly*, Robbins, R. J., and Melmed, S., Eds., Plenum Press, New York, 1987, 45.
44. **Amiel, S. A., Sherwin, R. S., Hintz, R. L., Gertner, J. M., Press, C. M., and Tamborlane, W. V.,** Effects of diabetes and its control on insulin-like growth factors in the patient with type I diabetes, *Diabetes*, 33, 1175, 1984.
45. **Brismar, K., Gutniak, M., Povoa, G., Werner, S., and Hall, K.,** Insulin regulates the 35 kDa IGF binding protein in patients with diabetes mellitus, *J. Endocrinol. Invest.*, 11, 599, 1988.
46. **Suikkari, A. M., Sane, T., Seppala, M., Jarvinen, H. Y., Karonen, S. L., and Koivist, V. A.,** Prolonged exercise increases serum insulin-like growth factor concentrations, *J. Clin. Endocrinol. Metab.*, 68, 141, 1989.
47. **Suikkari, A. M., Koivisto, V. A., Rutanen, E. M., Yki-Jarvinen, H., Karonen, S. L., and Seppala, M.,** Insulin regulates the serum levels of low molecular weight insulin-like growth factor-binding protein, *J. Clin. Endocrinol. Metab.*, 66, 266, 1988.
48. **Daughaday, W. H. and Deuel, T. F.,** Tumor secretion of growth factors, *Endocrinol. Metab. Clin. N. Am.*, 20, 539, 1991.
49. **Reeve, J. G., Brinkman, A., Hughes, S., Mitchell, J., Schwander, J., and Bleehen, N. M.,** Expression of insulin-like growth factor (IGF) and IGF-binding protein genes in human lung tumor cell lines, JNCI, 84, 628, 1992.
50. **Zapf, J., Futo, E., Peter, M., and Froesch, E. R.,** Can "big" insulin-like growth factor II in serum of tumor patients account for the development of extrapancreatic tumor hypoglycemia?, *J. Clin. Invest.*, 90, 2574, 1992.
51. **Schneid, H., Seurin, D., Noguiez, P., and Le Bouc, Y.,** Abnormalities of insulin-like growth factor (IGF-I and IGF-II) genes in human tumor tissue, *Growth Reg.*, 2, 45, 1992.
52. **Pekonen, F., Nyman, T., Ilvesmaki, V., and Partanen, S.,** Insulin-like growth factor binding proteins in human breast cancer tissue, *Cancer Res.*, 52, 5204, 1992.

53. **Zapf, J., Schmid, C., Guler, H. P., Waldvogel, M., Hauri, C., Futo, E., Hossenlopp, P., Binoux, M., and Froesch, E. R.,** Regulation of binding proteins for insulin-like growth factors (IGF) in humans. Increased expression of IGF binding protein 2 during IGF I treatment of healthy adults and in patients with extrapancreatic tumor hypoglycemia, *J. Clin. Invest.*, 86, 952, 1990.
54. **Cohen, P., Peehl, D. M., Stamey, T. A., Wilson, K., Clemmons, D. R., and Rosenfeld, R. G.,** Elevated levels of insulin-like growth factor binding protein-2 in the serum of prostate cancer patients, *J. Clin. Endocrinol. Metab.*, 76, 830, 1993.

CHAPTER 5

TRANSGENIC MICE IN THE STUDY OF GROWTH HORMONE AND THE INSULIN-LIKE GROWTH FACTORS

A. Joseph D'Ercole

TABLE OF CONTENTS

I. GROWTH HORMONE OVEREXPRESSION IN TRANSGENIC MICE

Studies employing transgenic (Tg) mice have served to clarify many of the actions of growth hormone (GH) and to dissect them from those actions mediated by insulin-like growth factor-I. They also have established the critical importance of the IGFs during development.

0-8493-8384-6/95/$0.00+$.50

A. GROWTH

Mice overexpressing rat (r) and human (h) GH were among the first transgenic mice generated.[1,2] Each of these lines of mice ectopically express a fusion gene driven by the mouse metallothionein-I (mMT-I) promoter. Using the same promoter, Tg mice have been created that also express bovine (b), ovine (o), porcine (p) and mouse GH.[3] All Tg mice overexpressing native GH molecules, including those that overexpress GH as a result of growth hormone releasing hormone (GHRH) overexpression, exhibit marked overgrowth.[4,5] They usually attain body weights that are about twice those of normal mice. The increase in body weight is accompanied by significant increases in skeletal growth, and the increase in some bone dimensions is as much as 32%. Interestingly, sexual dimorphism in body size is preserved in some lines of GH Tg mice (e.g., MT bGH), but not in others.[3] The reasons for these differences are not apparent, but may be related to the species of GH that is overexpressed.

The increases in somatic growth are not accompanied by similar increases in the size of all organs. In other words, the growth of each organ in these Tg animals is not proportional to the increase in body weight.[6] In general, the size of the kidney, liver, and spleen is disproportionally large. The growth of the heart and lung is proportional to somatic growth, and no discernable increase occurs in brain size. These findings indicate that GH has more growth-promoting effects on some tissues than on others.

Both the somatic and organ overgrowth resulting from GH overexpression occur postnatally between 3 and 13 weeks of age, despite increased GH expression from the later portion of *in utero* life. Thus, early growth of GH Tg mice is similar to that of normal mice, and the acceleration in growth rates occurs after weanling and persists for 2 to 3 months, a period of accelerated growth longer than the norm. The latter findings are consistent with evidence that GH is not crucial to prenatal growth, that GH actions temporally correlate with expression of GH receptors and, in turn, with the developmental increase in IGF-I expression that occurs in normal mice. In fact, an increase in hepatic IGF-I messenger RNA (mRNA) and in serum IGF-I concentrations was shown to precede the acceleration of growth in GH Tg mice.[7] Furthermore, the degree of GH overexpression, as measured by serum GH levels, does not correlate with the magnitude of overgrowth.

B. FERTILITY

Fertility is reduced in most lines of GH Tg mice.[3] For example, MT hGH Tg females are sterile, while males of this line exhibit reduced fertility. Similar findings have been observed in MT rGH and bGH mice. In contrast, Tg mice expressing MT oGH, as well as Tg mice carrying a hGH gene driven by the phosphoenolpyruvate carboxykinase promoter (PEPCK) and those overexpressing endogenous GH because of GH releasing hormone (GHRH) overexpression, exhibit normal or near-normal fertility.[4,8] It seems possible that the ecotopic site(s) of GH overexpression may influence fertility, although the influence of the different GH species expressed cannot be excluded. In MT hGH Tg mice altered gonadotropin secretion has been observed and this could account for the changes in fertility in these Tg mice.[9]

C. PATHOLOGY

Several abnormalities of serum chemistries have been noted in GH Tg mice. Serum IGF-I is elevated two- to threefold in all lines that were evaluated, but these elevations do not correlate with the elevation in serum GH. High serum insulin levels also were observed. Because the increase in insulin is not accompanied by alterations in serum glucose, these Tg mice can be considered insulin resistant. Consistent with insulin resistance, MT bGH Tg mice were found to have down-regulated hepatic insulin receptors.[10] Elevated serum cholesterols, but normal triglycerides, were found in all GH Tg mice so studied.

Growth hormone Tg mice exhibit significant histopathology.[3,11] Virtually all GH Tg lines develop severe renal pathology, which is the likely reason for their reduced life spans. The renal pathology is manifested by glomerular enlargement and progressive glomerulosclerosis, leading to end stage renal disease marked by nephron atrophy and tubular cystic lesions. The livers of GH Tg mice are usually markedly enlarged and characterized by centrilobular, hepatocellular, and hepatonuclear hypertrophy. Hepatic tumors also occur in GH Tg mice with increased frequency. Mammary adenocarcinoma reportedly occurred in hGH Tg mice.[12] Splenic congestion and red cell hematopoiesis are another common pathologic feature of GH Tg mice. Prins et al.[13] reported stromal hypertrophy of the seminal vesical in MT hGH Tg mice. Because similar changes have not been observed in MT bGH Tg mice, the lactogenic activity of hGH is thought to be responsible for these changes. The lactogenic activity is further supported by the findings that Tg hGH created with a fusion gene driven by the hydroxymethylglutaryl coenzyme-A reductase promoter have precocious mammary gland development and that virgin PEPCK hGH Tg female mice can produce milk and successfully rear foster litters.[8,14] Other abnormal histologic findings include an increased frequency of enlarged thyroid follicles and cardiovascular abnormalities.[15,16]

II. TRANSGENIC MICE WITH ABLATED GROWTH HORMONE EXPRESSION

Mice with pituitary somatotroph deficiency were created via the expression of a transgene that fuses the rGH promoter to the gene for diphtheria toxin.[17] Expression of this intracellular toxin directed by this cell-specific promoter results in the death of somatotrophs, and in turn absent GH expression and a GH-deficient phenotype. Using another cell ablation strategy, a common cellular lineage for pituitary somatotrophs and lactotrophs was shown using Tg mice that express the herpesvirus-1 thymidine kinase gene under the control of either the GH or prolactin promoters.[18] When the pituitary cells in these Tg mice are exposed to certain synthetic nucleoside analogues, the toxic metabolites generated by thymidine kinase ablate the expressing cells. Mice exposed to these agents, whose transgene is under the control of the GH promoter, are dwarfed and have neither somatotrophs nor lactotrophs, while those bearing the prolactin promoter-driven transgene exhibit normal pituitary glands and are of normal size.

III. TRANSGENIC MICE EXPRESSING MUTANT AND ALTERNATIVE FORMS OF GROWTH HORMONE

Transgenic mice were generated to elucidate the amino acid sequences of the GH molecule that are important to bioactivity by evaluating the growth-promoting actions of a variety of mutant bGH molecules encoded by transgenes.[19–22] To date these studies have been directed at the relationship between the structure and function of the third α-helix of bGH. Creation of an idealized amphipilic α-helix, generated by making three amino acid substitutions (glutamate-117 to leucine, glycine-119 to arginine, and alanine-122 to aspartate), resulted in a bGH molecule with normal receptor binding affinity. When expressed in Tg mice, it produced a dwarf phenotype, likely due to competitive antagonism with endogenous GH and subsequent inhibition of dimerization of the GH receptor.[19,22] A single amino acid substitution at amino acid 122 also results in dwarfism when expressed in Tg mice.[20] In other studies single amino acid substitutions at positions 114, 118, and 121 of the α-helix decreased secretion of bGH in transfected cultured cells, and in Tg mice, the mutated bGH molecules had an altered intracellular location.[21] The latter Tg mice also did not overgrow despite the presence of immunoreactive bGH in their serum.

Transgenic mice also were used in an attempt to distinguish the effects of the principal 22 kDa hGH from those of the naturally occurring, truncated 20-kDa form.[23] Transgenic mice expressing each form exhibited similar overgrowth and liver histopathology. The 22-kDa Tg mice, however, exhibited hyperalbuminemia and hypercholesterolemia and lower serum IGF-I levels than the 20-kDa Tg mice.

IV. TRANSGENIC MICE WITH ALTERED EXPRESSION OF GROWTH HORMONE REGULATORY PEPTIDES

The capacity of GHRH to stimulate pituitary hyperplasia, primarily somatotroph hyperplasia, and excess GH secretion was demonstrated using Tg mice that overexpress this peptide.[4,5] Growth hormone releasing hormone Tg mice were used to demonstrate that endogenous GH overexpression stimulates somatostatin expression in the hypothalamus.[24] In contrast, Tg mice with ablated somatotrophs exert relatively little influence on somatostatin expression (a 40% reduction in somatostatin mRNA). Transgenic mice expressing a MT-I somatostatin fusion gene were also made.[25] The most active site of expression of this transgene is the pituitary (predominantly by gonadotrophs). Despite this pituitary expression and high blood levels of somatostatin, the growth of these Tg mice is normal. The reasons for the lack of growth retardation are not clear.

In other Tg mice the importance of G proteins in the signaling pathways of GHRH was shown by inducing similar pituitary hyperplasia with a GH promoter-directed gene that expresses the intracellular cholera toxin, a potent stimulator of G_S.[26] A role of cyclic adenosine monophosphate (cAMP) in this signaling pathway was demonstrated in Tg mice that express a mutant cAMP response element binding protein.[27] The mutant protein cannot be phosphorylated, and therefore is inactive. When overexpressed in the pituitary the mutant binding protein competes with the native one. Because transcription of cAMP-regulated factors does not occur, atrophied pituitary glands result.

V. INSULIN-LIKE GROWTH FACTOR-I TRANSGENIC MICE

If IGF-I mediates all the growth-promoting actions of GH, Tg IGF-I overexpressing mice might be expected to exhibit growth that is identical to that of GH overexpressing mice. To directly test this hypothesis, Mathews et al.[28] constructed a hIGF-I transgene using MT-I as a promoter. A hIGF-IA cDNA was used in this construct, because the IGF-I gene is too large (> 75 kb in humans) to be incorporated in a fusion gene.[29] The IGF-I transgene thus differed from GH fusion genes used to make Tg mice in that the latter used GH genes including intronic sequences (which appear to effect more efficient expression in Tg mice).[30] Because the IGF-I gene encodes several putative signal peptides and the signal peptide that is optimal for secretion has not been identified, the IGF-I transgene was constructed with sequences coding the signal peptide of somatostatin, a sequence known to be functional in Tg mice. Only a single founder that exhibited a significant increase in serum IGF-I was identified. In subsequent matings female Tg mice were found to be infertile (as are most lines of GH overexpressing mice). Therefore, IGF-I Tg mice were maintained by breeding male Tg mice to normal mice, and all Tg mice are hemizygous for the transgene.

A. GROWTH IN INSULIN-LIKE GROWTH FACTOR-I TRANSGENIC MICE

Insulin-like growth factor-I Tg mice become significantly larger than their nontransgenic siblings after 6 weeks of age. The trend toward overgrowth, however, can be observed by about 4 weeks of life, a time similar to when overgrowth occurs in GH overexpressing mice.[31] At 8 weeks of age, these mice exhibit a 30% increase (approx.) in body weight. Of the organs studied at this age, brain, carcass, kidney, pancreas, and spleen exhibited significant overgrowth. Because DNA content was similarly increased in the overgrown organs, hyperplasia is likely to account for the increases in organ size. While the carcass size was only increased by about 21%, this increase accounted for about 75% of the increase in body weight, indicating that IGF-I overexpression has significant effects on skeletal, connective, adipose tissue, and/or muscle growth. However, there was not a convincing increase in linear or skeletal growth, as assessed by the measurement of tail lengths, cumulative tibial width growth, or long bone length.

B. DIFFERENT PHENOTYPES IN GROWTH HORMONE AND INSULIN-LIKE GROWTH FACTOR-I TRANSGENIC MICE

1. Growth

The most dramatic differences between IGF-I and GH Tg mice were in the sizes of the liver and brain. The brains of IGF-I transgenic mice weighed 50% more than those of their normal littermates and contained 21% more DNA, while brains of GH Tg mice are not significantly increased in size. Liver size, however, is dramatically increased in GH Tg mice but not overgrown in IGF-I Tg mice.

2. Pathology

Pathologic differences also occurred between these lines of mice. Insulin-like growth factor-I mice showed only minimal pathology.[11] They exhibited a modest enlargement of renal glomerular size, splenic congestion and hematopoiesis, and some difficult to define dermal abnormalities that appeared to represent adipose thickening and collagen bundle disruption. Unlike GH Tg mice, IGF-I Tg mice have suppressed serum insulin levels, although their serum glucose concentrations also are normal.[11] Also in contrast to GH Tg mice, IGF-I Tg mice exhibit elevated serum triglycerides, but normal serum cholesterol.

3. Analysis of Mechanisms

While the overgrowth of IGF-I Tg mice demonstrated that IGF-I can stimulate growth *in vivo,* the differences in growth phenotypes between these and GH Tg animals does not confirm the somatomedin hypothesis in its simplest interpretation, i.e., IGF-I mediates all the growth promoting actions of GH. A number of issues relevant to the physiology of GH and IGF-I, however, make it difficult to dismiss the somatomedin hypothesis on the basis of the differing phenotypes among these two transgenic models.

Growth hormone acts by endocrine mechanisms of action, i.e., it is synthesized and secreted in a single organ, the pituitary, and is transported by the circulation to its many distant sites of action. The ectopic expression of GH, as occurs in transgenic mice (liver, etc.), therefore, may not be a critical factor in determining its actions, as long as the ectopic expression results in a significant elevation of blood-borne GH. In other words, the consequences of GH overexpression will accrue if circulating GH levels are sufficiently high. The mode of action of IGF-I is more complex.[29,32,33] While IGF-I normally circulates in concentrations that appear sufficiently high to stimulate biologic actions, it also is synthesized in most organs and tissues. Therefore, IGF-I almost certainly acts locally on or near its cells of synthesis, in addition to its probable endocrine actions. The precise cellular sites of IGF-I expression in Tg mice may be key determinants of its actions. Furthermore, the specific signal and trailer peptide portions of the IGF-I precursor synthesized may be important in determining its action. If the aforementioned factors are critical to IGF-I actions, then the IGF-I expressed by the transgene used may not be capable of replicating the actions of endogenous IGF-I in every organ where it is expressed, because either the cell of expression or precursor expressed are not optimal. On the other hand, transgene GH would be expected to stimulate the appropriate endogenous IGF-I precursor in its usual cellular site of expression. Thus, actions of GH that are mediated by IGF-I may be manifested in GH Tg animals but not necessarily occurring in this line of IGF-I Tg mice.

The IGF-I transgene, as assessed with a human specific solution hybridization assay, was expressed in liver, pancreas, lung, kidney, spleen, and intestine, and, as assessed by Northern analysis and *in situ* hybridization, in brain. Extracted immunoreactive IGF-I was also increased in each of these organs, and the magnitude of RNA and peptide expression were in general agreement. The magnitude of IGF-I transgene expression in specific organs, however, does not correlate well with the degree of overgrowth. For example, pancreas, which showed the highest IGF-I transgene expression, was also proportionally the most overgrown, but brain and spleen, which exhibited proportional overgrowth, did not exhibit high transgene expression. The latter tissues may be especially sensitive to the effects of IGF-I or the IGF-IA precursor expressed by the transgene. In the case of spleen it seems possible that blood-borne IGF-I is the stimulator of the marked overgrowth. Other alternative explanations, however, seem possible.

Serum IGF-I concentrations were increased by about 50% in IGF-I Tg mice, and thus, endocrine effects of IGF-I could have contributed to the overgrowth. This circulating IGF-I likely exerted negative feedback regulation on the pituitary, because pituitary GH mRNA was depressed by about 60%.[31] No change took place in somatostatin mRNA, as judged by *in situ* hybridization in paraventricular nuclei, and therefore, a direct effect on somatotrophs seems likely. The suppressed GH secretion in turn resulted in a depression of endogenous hepatic mouse IGF-I expression to about 30% of normal. Presumably, endogenous IGF-I expression also is depressed in other GH responsive tissues. Given this GH dependence of endogenous IGF-I expression in many tissues, organ-specific overgrowth in IGF-I Tg mice may depend to a great extent upon whether transgene IGF-I is expressed at specific tissue locations and in sufficient quantity.

The degree of IGF-I overexpression also may explain some of the differences in growth between the GH and IGF-I Tg mice. Growth hormone Tg mice express higher levels of hepatic IGF-I mRNA than do IGF-I mice. The GH Tg mouse lines also generally have higher IGF-I serum concentrations. In addition, hepatic IGF-I mRNA expression and serum IGF-I concentrations in GH Tg mice correlate better with body size than does GH expression.[7] Finally, as judged by blood IGF-I levels, the degree of IGF-I overexpression is minimal. While being significantly elevated on average, serum IGF-I concentrations are within the range of normal in many IGF-I transgenic mice. Any endocrine effects of IGF-I, therefore, would not be expected to result in dramatic overgrowth.

C. TRANSGENIC MICE EXPRESSING INSULIN-LIKE GROWTH FACTOR-I, BUT NOT GROWTH HORMONE

The modest IGF-I overexpression in the IGF-I Tg line, confounded by the persistence of GH expression (albeit suppressed), made it seem reasonable to ask whether the transgene IGF-I expression exhibited in this line could replicate normal, or stimulate excessive growth, in the face of GH deficiency. To create this situation, hemizygous Tg mice with isolated GH deficiency, i.e., Tg mice carrying the transgene that fuses the rGH promoter to the gene encoding the A chain of diphtheria toxin (DT; i.e., DT mice), were crossed with hemizygous IGF-I Tg mice.[34] The mice of this cross that carry both transgenes would be expected to express IGF-I in the absence of GH.

1. Growth

Both body weight and tail length of IGF-I expressing, GH-deficient mice became significantly greater than those of GH-deficient, DT mice by 30 d of age; however, an acceleration of growth was apparent as early as 1 to 2 weeks of age. Equally important, the weight and length of the GH-deficient mice expressing IGF-I could not be distinguished from that of normal mice. At the same time, the IGF-I Tg mice overgrew. The overgrowth, however, was not significant until after 4 weeks of age, and as in the initial study, the tail lengths of these mice was not significantly greater than those of normal mice. The tail growth was proven to be a reflection of skeletal growth because the cumulative increase in tibial epiphyseal width was much greater in IGF-I expressing, GH-deficient mice than in dwarfs and similar to that of normals and IGF-I Tg mice. These studies demonstrated that IGF-I expression by itself can replicate normal somatic growth, if one defines it as growth of body weight and skeletal growth.

2. Analysis of Mechanisms

These studies also provided evidence to explain the modest overgrowth of the IGF-I Tg animals as compared to that of GH Tg mice. The data argue that expression of the IGF-I transgene, while sufficient to stimulate normal growth, is not sufficient to result in overgrowth except in organs especially sensitive to IGF-I or in those where transgene expression is especially high (discussed above). Analysis of endogenous and transgene hepatic RNA abundance (measured using species specific sequences in solution hybridization assays) in the mice of this cross forms part of this explanation. Endogenous IGF-I mRNA was about 8% of normal in both groups of GH-deficient mice, the same degree of suppression that was observed after hypophysectomy in rats.[35] As previously shown, endogenous IGF-I mRNA was suppressed in the IGF-I Tg mice, presumably resulting from suppressed GH expression and secretion.[31] The IGF-I transgene, however, was found to be less well expressed in the mice carrying only the IGF-I transgene than in the IGF-I expressing, GH-deficient mice, suggesting that endogenous IGF-I expression or GH expression had a negative influence on the expression of the IGF-I transgene. Furthermore, the data suggest that IGF-I expression, regardless of the gene directing its synthesis, is regulated and provides a mechanism to limit overgrowth. In addition, when the abundance of endogenous and transgene IGF-I transcripts is totaled, it is apparent that the expression of liver IGF-I transcripts was not increased above normal in the IGF-I Tg mice. The IGF-I transgenic mice, thus, did not overexpress IGF-I in liver. These findings suggest that much of the overgrowth observed in IGF-I Tg mice results from IGF-I transgene expression in other tissues.

Serum IGF-I concentrations in these mice were consistent with the transgene providing near-normal IGF-I expression, rather than overexpression. Serum IGF-I in IGF-I expressing, GH-deficient mice were near those of normal mice (1.29 ± 0.15 vs. 1.00 ± 0.22 U/ml; mean ± SD), and were likely derived almost entirely from the transgene, because serum IGF-I in GH-deficient mice is only 0.04 ± 0.02 U/ml. The serum IGF-I in the IGF-I Tg mice (1.57 ± 0.33 U/ml) also was likely derived in large part from the transgene because endogenous IGF-I RNA expression was suppressed. Finally, serum IGF-I levels correlated well with both body weight and tail length in the mice of this cross. If serum IGF-I can be taken as a valid index of IGF-I expression, then the growth of these mice depended upon the degree of IGF-I expression, as implied by the somatomedin hypothesis.

The growth of some organs studied (lung, pancreas, and duodenum) paralleled the changes in body weight among the four genotypes. In other words, the proportional size of these organs, calculated as a percentage of body weight, remained constant. In these organs weights were lowest in GH-deficient mice, similar in the normals and in mice that express IGF-I in the absence of GH, and modestly enlarged in IGF-I Tg mice. These findings suggested that as with somatic growth, IGF-I mediates the growth promoting effects of GH.

The pattern of growth was different in the liver and brain. Liver weights were lowest in GH-deficient mice. The liver also comprised a lower percentage of body weight in these mice, indicating that either GH, or the IGF-I that it induces, is important to growth. When IGF-I is expressed in the DT mice, liver weight increased, but not to normal size. Also, this increase was not proportional to the increase in body weight; i.e., its percentage of body weight was similar to that of GH-deficient mice. The IGF-I Tg mice had increased liver weights that represented a normal percentage of body size. To interpret these data two other findings must be recalled: (1) IGF-I Tg mice express GH, and thus, it is not possible to exclude a role for GH in the increase in liver size in IGF-I Tg mice, and (2) GH Tg mice exhibit marked liver overgrowth. It seems, therefore, that GH has a special role in liver growth that is independent of IGF-I.

The changes in brain size among the mice of this cross were the most dramatic of those observed. Brain size was different in mice of each genotype. Both groups of mice that express the IGF-I transgene had much bigger brains than their counterparts that did not express this transgene; i.e., weights were greater in the IGF-I expressing, GH-deficient mice than in the GH-deficient mice, and greater in the IGF-I Tg mice than in the normal mice. The brain comprised about the same percentage of body weight in each group of mice. In interpreting the latter finding it must be kept in mind that the majority of brain growth occurs early in development, and thus postnatal brain growth does not parallel somatic growth in normal mice. Each of these mice had normal brain sizes at birth. In addition, GH Tg mice exhibit little or no increase in brain size. The data, therefore, support the conclusion that IGF-I has a special role in brain growth and that this function is independent of that of GH. *In situ* hybridization studies have shown that the mRNA for the IGF-I transgene localizes to the choroid plexus and ependyma, and therefore, provides evidence that the IGF-I synthesized in the brain acts locally to stimulate the increases in brain size.

3. Insulin-Like Growth Factor Binding Proteins

The mice from this cross also proved useful in dissecting the role of GH and IGF-I in the regulation of IGF binding proteins (IGFBPs).[36] Serum IGFBP-3, as assessed by ligand blot analysis, was marked by depressed in GH-deficient mice as compared to normals. It was restored toward normal, however, in IGF-I expressing, GH-deficient mice, and was increased 2.1-fold in IGF-I Tg mice. These findings indicate that IGFBP-3 is under the control of IGF-I, rather than GH.

The so-called GH-dependent, 140-kDa, serum IGF-binding protein complex is composed of IGF-I or IGF-II, IGFBP-3, and another nonbinding subunit, called the acid labile subunit.[37] This complex could not be detected in the serum of IGF-I expressing, GH-deficient mice, despite near-normal IGF-I and IGFBP-3 concentrations. The 140 kDa complex, however, was present in IGF-I Tg mouse sera. These findings, therefore, indicated that the acid labile subunit in the 140 kDa complex is GH dependent. Furthermore, both IGF-I and GH (e.g., as expressed in IGF-I Tg mice) are necessary for the formation of this IGF-binding protein complex, which accounts for most of the IGFs in serum. Comparison of sera from these mice also showed that IGFBP-2 (evaluated by immunoblot) was increased by IGF-I and suppressed by GH, and that IGFBP-4 may be stimulated modestly by IGF-I. Because IGFBPs are known to modulate the actions of IGF-I, their complex regulation by both IGF-I and GH, as well as by other factors, points to another level of growth control.[38] Differences in IGFBP expression could provide alternative explanations accounting for the differing phenotypes in GH and IGF-I transgenic mice.

VI. TRANSGENIC MICE WITH DISRUPTED INSULIN-LIKE GROWTH FACTOR AND INSULIN-LIKE GROWTH FACTOR RECEPTOR GENES

The use of homologous recombination to substitute endogenous genes with artificial fusion genes provided a strategy to disrupt a specific gene, and thus, provides models to deduce a gene's function by observing the consequences of its absence. Pluripotent embryonic stem cells (ESC) are transfected with fusion genes, and subsequently the cells undergoing homologous recombination are selected and amplified. These selected ESC are then injected into blastocysts that are transferred to pseudopregnant mice for *in utero* development. Mosaic mice with both native and ESC-derived cells result. Mice with germ lines composed of ESC-derived cells can then be used to develop lines hemi- or homozygous for the recombinant transgene. Using this technique, DiChiara and co-workers[39] created mice in whom the IGF-II gene was disrupted. They showed that marked fetal growth retardation (a 40% decrease in fetal size) occurs in mice bearing a hemizygous disruption of the IGF-II gene. The reduced size of these mice occurs before day 17 of gestation (mouse gestation is 19 d), a finding consistent with the abundant expression of IGF-II before this time and its proposed role in the stimulation of fetal growth. Despite their small size at birth, these Tg mice grow at normal rates postnatally, although they do not exhibit ''catch-up'' growth, and are fertile.

Another important finding from the study of these Tg IGF-II deleted mice was the observation that the IGF-II gene is paternally imprinted; i.e., only the paternally derived gene is expressed in offspring. DeChiara et al.[40] observed that all the offspring of mothers hemizygous for the disrupted gene were normal, while the offspring of Tg male mice were runted if they carried the disrupted gene. Interestingly, the type II IGF/mannose-6 phosphate receptor is maternally imprinted.[41,42] This finding comes from studies of the Tme and related mouse lines. These naturally occurring mouse lines have genomic deletions that include a complete deletion of the gene for the type II IGF receptor. When the deletion is maternally inherited, the affected offspring are edematous and die at about 15 d gestation. The pathology of these mice, however, may be directly due to the effects of an overabundance of IGF-II.[43] Filson et al.[43] crossed T^{hp} mice lacking the type II IGF receptor with the Tg mice carrying the IGF-II disruption and found that the mice bearing both mutants did not die *in utero*; thus, this finding strongly indicates that a lack of IGF-II expression rescues these mice. This interpretation is consistent with the known role of the type II IGF receptor in cycling IGF-II, as well as proteins containing mannose-6 phosphate moieties, to lysosomes where they can be degraded. In other words, it appears that an excess of IGF-II may be toxic to the fetus and that a major function of the type II IGF receptor is to regulate IGF-II concentration by promoting its degradation.

The effects of disruptions of the IGF-I and the type I IGF receptor genes were recently reported.[44,45] Efstratiadis' group created mice with deletions of each of these genes. While mice with a hemizygous disrupted IGF-I gene appear normal, Tg mice with a homozygous mutant IGF-I gene exhibit growth retardation similar to those who lack IGF-II (40% weight retarded). When the Tg mice expressing no IGF-I, i.e. (homozygous IGF-I disrupted mice) are bred with mice of their original strain (i.e., inbred), about 90% die soon after birth. When bred into different genetic backgrounds, i.e., into mouse strains different from the founder mice, survival increases but growth is poor throughout development, with adult size being only about 30% of normal. These mice also are infertile.

Transgenic mice with a heterozygous disruption of the type I IGF receptor gene are phenotypically normal and express normal quantities of the type I receptor mRNA and functional receptors.[44,45] Transgenic mice with a homozygous disruption of this gene are markedly growth retarded (55% decrease in fetal size) and invariably die after birth. The pathology exhibited is similar whether the mice lack IGF-I or the type I receptor, and is marked by generalized organ hypoplasia, delayed ossification, and abnormalities in the central nervous system and epidermis.

Crosses of these mice bearing disruptions of IGF and type I receptor genes also provided valuable information.[44,45] An absence of IGF-I expression does not worsen the phenotype seen in mice lacking type I IGF receptor expression (55% growth retardation and invariable death at birth), indicating that all the effects of absent IGF-I expression are mediated by the type I receptor. Mice bearing homozygous disruptions of both the IGF-II and type I receptor genes, however, are more growth retarded at birth than mice with a homozygous deletion of only one of these genes, being about 30% of normal size. Because the type II receptor does not appear to mediate the growth-promoting actions of IGFs, these data suggest that IGF-II also acts through another, as yet unidentified, receptor in addition to the type I receptor. Analysis of the growth of these mice throughout gestation suggests that this unidentified receptor is functional after 13.5 d of gestation, because it is after this time in gestation that the growth retardation worsens in the mice with an absence of both IGF-II and type I receptor expression. In addition, this unidentified receptor appears to be essential to placental growth because the growth of this organ is normal in mice with an absence of either IGF-I or type I receptor expression (or both), but retarded in mice that do not express IGF-II. Placental growth retardation also is not worsened in mice that lack type I receptor expression in addition to an absence of IGF-II expression. In total, these elegant studies prove that both IGF-I and IGF-II are necessary for normal growth *in utero,* and that the actions of both IGFs in the fetus are primarily mediated by the type I IGF receptor, although another receptor is important to the actions of IGF-II, especially in the placenta.

REFERENCES

1. **Palmiter, R. D., Brinster, R. L., Hammer, R. E., Trumbauer, M. E., Rosenfeld, M. G., Birnberg, N. C., and Evans, R. M.,** Dramatic growth of mice that develop from eggs microinjected with metallothionein-growth hormone fusion genes, *Nature,* 300, 611, 1982.
2. **Palmiter, R. D., Norstedt, G., Gelinas, R. E., Hammer, R. E., and Brinster, R. L.,** Metallothionein-human GH fusion genes stimulate growth of mice, *Science,* 222, 809, 1983.
3. **Wanke, R., Wolf, E., Hermanns, W., Folger, S., Buchmüller, T., and Brem, G.,** The GH-transgenic mouse as an experimental model for growth research: clinical and pathological studies, *Horm. Res.,* 37 (Suppl. 3), 74, 1992.
4. **Hammer, R. E., Brinster, R. L., Rosenfeld, M. G., Evans, R. M., and Mayo, K. E.,** Expression of human growth hormone-releasing factor in transgenic mice results in increased somatic growth, *Nature,* 315, 413, 1985.

5. **Mayo, K. E., Hammer, R. E., Swanson, L. W., Brinster, R. L., Rosenfeld, M. G., and Evans, R. M.,** Dramatic pituitary hyperplasia in transgenic mice expressing a human growth hormone-releasing factor gene, *Mol, Endocrinol,* 2, 606, 1988.
6. **Shea, B. T., Hammer, R. E., and Brinster, R. L.,** Growth allometry of organs in giant transgenic mice, *Endocrinology,* 121, 1924, 1987.
7. **Mathews, L. S., Hammer, R. E., Brinster, R. L., and Palmiter, R. D.,** Expression of insulin-like growth factor I in transgenic mice with elevated levels of growth hormone is correlated with growth, *Endocrinology,* 123, 433, 1988.
8. **Milton, S., Cecim, M., Li, Y. S., Yun, J. S., Wagner, T. E., and Bartke, A.,** Transgenic female mice with high human growth hormone levels are fertile and capable of normal lactation without having been pregnant, *Endocrinology,* 131, 536, 1992.
9. **Chandrashekar, V., Bartke, A., and Wagner, T. E.,** Neuroendocrine function in adult female transgenic mice expressing the human growth hormone gene, *Endocrinology,* 130, 1802, 1992.
10. **Balbis, A., Dellacha, J. M., Calandra, R. S., Bartke, A., and Turyn, D.,** Down regulation of masked and unmasked insulin receptors in the liver of transgenic mice expressing bovine growth hormone gene, *Life Sci.,* 51, 771, 1992.
11. **Quaife, C. J., Mathews, L. S., Pinkert, C. A., Hammer, R. E., Brinster, R. L., and Palmiter, R. D.,** Histopathology associated with elevated levels of growth hormone and insulin-like growth factor I in transgenic mice, *Endocrinology,* 124, 40, 1989.
12. **Törnell, J., Rymo, L., and Isaksson, O. G. P.,** Induction of mammary adenocarcinomas in metallothionein promoter-human growth hormone transgenic mice, *Int. J. Cancer,* 49, 114, 1991.
13. **Prins, G. S., Cecim, M., Birch, L., Wagner, T. E., and Bartke, A.,** Growth response and androgen receptor expression in seminal vesicles from aging transgenic mice expressing human or bovine growth hormone genes, *Endocrinology,* 131, 2016, 1992.
14. **Bchini, O., Andres, A. C., Schubaur, B., Mehtali, M., LeMeur, M., Lathe, R., and Gerlinger, P.,** Precocious mammary gland development and milk protein synthesis in transgenic mice ubiquitously expressing human growth hormone, *Endocrinology,* 128, 539, 1991.
15. **Mayerhofer, A., Easterly, S., Amador, A. G., Gher, J., Bartke, A., Yun, J., and Wagner, T. E.,** Studies on the thyroid in transgenic mice expressing the genes for human and bovine growth hormone, *Experientia,* 46, 1043, 1990.
16. **Dilley, R. J. and Schwartz, S. M.,** Vascular remodeling in the growth hormone transgenic mouse, *Circ. Res.,* 65, 1233, 1989.
17. **Behringer, R. R., Mathews, L. S., Palmiter, R. D., and Brinster, R. L.,** Dwarf mice produced by genetic ablation of growth hormone-expressing cells, *Gene Dev.,* 2, 453, 1988.
18. **Borelli, E., Heyman, R. A., Arias, C., Sawchenko, P. E., and Evans, R. M.,** Transgenic mice with inducible dwarfism, *Nature,* 339, 538, 1989.
19. **Chen, W. Y., Wight, D. C., Wagner, T. E., and Kopchick, J. J.,** Expression of a mutated growth hormone gene suppresses growth of transgenic mice, *Proc. Natl. Acad. Sci. U.S.A.,* 87, 5061, 1990.
20. **Chen, W. Y., Wight, D. C., Mehta, B. V., Wagner, T. E., and Kopchick, J. J.,** Glycine 119 of bovine growth hormone is critical for growth-promoting activity, *Mol. Endocrinol.,* 5, 1845, 1991.
21. **Chen, W. Y., Wight, D. C., Chen, N. Y., Coleman, T. A., Wagner, T. E., and Kopchick, J. J.,** Mutations of the third α-helix of bovine growth hormone dramatically affect its intracellular distribution in vitro and growth enhancement in transgenic mice, *J. Biol. Chem.,* 266, 2252, 1991.
22. **Chen, W. Y., White, M. E., Wagner, T. E., and Kopchick, J. J.,** Functional antagonism between endogenous mouse growth hormone (GH) and a GH analog results in dwarf transgenic mice, *Endocrinology,* 129, 1402, 1991.

23. **Stewart, T. A., Clift, S., Pitts-Meek, S., Martin, L., Terrell, T. G., Liggitt, D., and Oakley, H.,** An evaluation of the functions of the 22-kilodalton (kDa), the 20-kDa, and the N-terminal polypeptide forms of human growth hormone using transgenic mice, *Endocrinology,* 130, 405, 1992.
24. **Hurley, D. L. and Phelps, C. J.,** Hypothalamic preprosomatostatin messenger ribonucleic acid expression in mice transgenic for excess or deficient endogenous growth hormone, *Endocrinology,* 130, 1809, 1992.
25. **Low, M. J., Hammer, R. E., Goodman, R. H., Habener, J. F., Palmiter, R. D., and Brinster, R. L.,** Tissue-specific posttranslational processing of pre-prosomatostatin encoded by a metallothionein-somatostatin fusion gene in transgenic mice, *Cell,* 41, 211, 1985.
26. **Burton, F. H., Hasel, K. W., Bloom, F. E., and Sutcliffe, J. G.,** Pituitary hyperplasia and gigantism in mice caused by a cholera toxin transgene, *Nature,* 350, 74, 1991.
27. **Struthers, R. S., Vale, W. W., Arias, C., Sawchenko, P. E., and Montminy, M. R.,** Somatotroph hypoplasia and dwarfism in transgenic mice expressing a non-phosphorylatable CREB mutant, *Nature,* 350, 622, 1991.
28. **Mathews, L. S., Hammer, R. E., Behringer, R. R., Brinster, R. L., and Palmiter, R. D.,** Transgenic mice as experimental models for elucidating the role of growth hormone and insulin-like growth factors for body growth, in *Growth Hormone: Basic and Clinical Aspects,* Isaksson, O., Binder, C., Hall, K., and Hokfelt, B., Eds., Exerpta Medica, Amsterdam, 1987, 373.
29. **Daughaday, W. H. and Rotwein, P.,** Insulin-like growth factors I and II. Peptide, messenger ribonucleic acid and gene structures, serum, and tissue concentrations, *Endocrinol. Rev.,* 10, 68, 1989.
30. **Palmiter, R. D., Sandgren, E. P., Avarbock, M. R., Allen, D. D., and Brinster, R. L.,** Heterologous introns can enhance expression of transgenes in mice, *Proc. Natl. Acad. Sci. U.S.A.,* 88, 478, 1991.
31. **Mathews, L. S., Hammer, R. E., Behringer, R. R., D'Ercole, A. J., Bell, G. I., Brinster, R. L., and Palmiter, R. D.,** Growth enhancement of transgenic mice expressing human insulin-like growth factor I, *Endocrinology,* 123, 2827, 1988.
32. **D'Ercole, A. J.,** The somatomedins/insulin-like growth factors, in *Clinical Paediatric Endocrinology,* Brook, C. G. D., Ed., Blackwell Scientific, Oxford, 1989, 74.
33. **Van Wyk, J. J. and Trippel, S. B.,** Endocrine, paracrine, and autocrine effects of the somatomedins/insulin-like growth factors, in *Growth Hormone: Basic and Clinical Aspects,* Isaksson, O., Binder, C., Hall, K., and Hokfelt, B., Eds., Elsevier, Amsterdam, 1987, 337.
34. **Drop, S. L. S., Kortleve, D. J., and Guyda, H. J.,** Isolation of a somatomedin-binding protein from preterm amniotic fluid. Development of a radioimmunoassay, *J. Clin. Endocrinol. Metab.,* 59, 899, 1984.
35. **Mathews, L. S., Norstedt, G., and Palmiter, R. D.,** Regulation of insulin-like growth factor I gene expression by growth hormone, *Proc. Natl. Acad. Sci. U.S.A,* 83, 9343, 1986.
36. **Camacho-Hubner, C., Clemmons, D. R., and D'Ercole, A. J.,** Regulation of insulin-like growth factor (IGF) binding proteins in transgenic mice with altered expression of growth hormone and IGF-I, *Endocrinology,* 129, 1201, 1991.
37. **Baxter, R. C. and Martin, J. L.,** Structure of the Mr 140,000 growth hormone-dependent insulin-like growth factor binding protein complex: determination by reconstitution and affinity-labeling, *Proc. Natl. Acad. Sci. U.S.A.,* 86, 6898, 1989.
38. **Clemmons, D. R.,** Insulin-like growth factor binding protein control secretion and mechanisms of action, *Adv. Exp. Med. Biol.,* 293, 113, 1991.
39. **DeChiara, T. M., Efstratiadis, A., and Robertson, E. J.,** A growth-deficiency phenotype in heterozygous mice carrying an insulin-like growth factor II gene disrupted by targeting, *Nature,* 345, 78, 1990.
40. **DeChiara, T. M., Robertson, E. J., and Efstratiadis, A.,** Parental imprinting of the mouse insulin-like growth factor II gene, *Cell,* 64, 849, 1991.

41. **Barlow, D. P., Stöger, R., Herrmann, B. G., Saito, K., and Schweifer, N.,** The mouse insulin-like growth factor type-2 receptor is imprinted and closely linked to the *Tme* locus, *Nature,* 349, 84, 1991.
42. **Stöger, R., Kubicka, P., Liu, C.-G., Kafri, T., Razin, A., Cedar, H., and Barlow, D. P.,** Maternal-specific methylation of the imprinted mouse *Igf2r* locus identifies the expressed locus as carrying the imprinting signal, *Cell,* 73, 61, 1993.
43. **Filson, A. J., Louvi, A., Efstratiadis, A., and Robertson, E. J.,** Rescue of the T-associated maternal effect in mice carrying null mutations in *Igf-2* and *Igf2r,* two reciprocally imprinted genes, *Development,* 118, 731, 1993.
44. **Liu, J. P., Baker, J., Perkins, A. S., Robertson, E. J., and Efstratiadis, A.,** Mice carrying null mutations of the genes encoding insulin-like growth factor I (Igf-1) and type 1 IGF receptor, *Cell,* 75, 59, 1993.
45. **Baker, J., Liu, J. P., Robertson, E. J., and Efstratiadis, A.,** Role of insulin-like growth factors in embryonic and postnatal growth, *Cell,* 75, 73, 1993.

Chapter 6

CLINICAL ASPECTS OF GROWTH HORMONE TREATMENT OF CHILDREN

Alan D. Rogol

TABLE OF CONTENTS

I. INTRODUCTION

Continued growth of an organism is generally considered a sign of health and well being. How a child compares with his peers can be derived from comparison of that child with the normal group represented on a standard growth chart. By definition, *normal* (physiologic) growth encompasses the 95% confidence interval for a specific group of subjects. Most children with a normal growth pattern, but who remain below the lower 2.5 percentile (approximately −2 SD) will also be otherwise normal; however, the further below the −2 SD level, the more likely that the child has a condition that is keeping him or her from reaching the genetically endowed adult height.

A. NORMAL GROWTH

Normal variants of growth (see below) were found in 82% of children with height less than the third percentile and in 50% of children whose height was

0-8493-8384-6/95/$0.00+$.50

>3 SD below the mean for age in a population study from Newcastle-upon-Tyne, England.[1] Even at a university referral service virtually one half of children evaluated for height standard deviation score below −2 SD were found to have a physiologic variant as the "cause" of their short stature.[2] On the other hand, growth failure at any point other than full maturity must be considered pathologic no matter where on the growth curve a particular child stands. Any single point on the growth chart is not very informative. When several growth points are plotted, however, it should become apparent whether the child's growth is average, a variant of normal or pathologic. The point at which a child is placed at any given time can be related to the height age, that age at which the child's height would be at the 50th percentile. The height age is obtained from the growth chart by drawing a line parallel to the chronological age axis from the child's plotted point to the 50th percentile. The height age is then determined along the chronological age scale at the intersection of the drawn line and the 50th percentile. It is an indication of the mean age of children of that given height. Most children who have abnormalities of growth will have retarded height age compared to chronological age, although a few will demonstrate accelerated growth.

The skeletal or dental maturation may be used to assess a child's developmental status. The interpretation of osseous maturation by the method of Greulich and Pyle[3] has proved to be convenient and reliable. A single radiograph of the left hand and wrist is obtained and the developmental maturation (bone age) is compared to that of children of normal stature using an atlas. Because girls are more developmentally advanced for any chronological age, separate standards exist for girls and boys.

The normal growth curve is characterized by rapid fetal growth with a peak intrauterine growth velocity at 4 months of gestational age followed by marked deceleration of growth after birth, and then a period of relatively slow but constant growth during childhood, and finally a rapid growth spurt at puberty, falling to zero velocity at epiphyseal closure.

The minimal normal growth velocity is generally accepted to be 5 cm/year, although a small percentage of normal children will grow at slower rates for varying periods. Size at birth is determined by intrauterine environment as well as by genetic (polygenic) factors. Between birth and 2 years of age infants make adjustments for these maternal factors and increase or decrease their growth velocity in relationship to the norm to reach their genetically determined growth potential.[4] After 2 years of age we find a tendency for the child to follow this growth channel. Any deviation from this percentile on the growth curve may indicate a pathologic process interfering with normal growth. A simple rule of thumb is that a child grows 10 in. (25 cm) in the first year, half that (5 in., 12.5 cm) in the second, and half that (2½ in., 6.4 cm) each year thereafter until puberty. An average newborn measures 20 in. (51 cm), therefore, an average 1 year old measures 30 in. (76 cm), a 2 year old 35 in. (89 cm), a 4 year old 40 in. (102 cm), and an 8 year old 50 in. (127 cm). The preadolescent nadir in growth rate is often exaggerated in teenagers with delayed puberty. The growth rate may decelerate to below 4 cm/year occurring just before the beginning of the adolescent growth spurt at the appropriate biological (skeletal) age, which may be several years behind chronologic age.

B. GROWTH DISORDERS

Taken together the criteria of height age, growth velocity, and bone age permit one to determine a child's growth pattern and development. They should then help to determine whether an individual child's growth is within the broad range of normal or pathologic. The above criteria permit classification of growth disorders into three distinct groups:[5]

- intrinsic short stature
- delayed growth
- attenuated growth

These patterns are shown graphically in Figure 1.

C. INTRINSIC SHORT STATURE

Intrinsic short stature is characterized by short stature, growth rate within normal limits, and no delay in skeletal development. Puberty occurs at the appropriate chronological age, resulting in short adult stature. Although this pattern most often reflects multiple genetic influences and is commonly called "familial short stature", it may also be the result of incomplete recovery from severe intrauterine growth retardation or mild skeletal dysplasia.

D. DELAYED GROWTH

Delayed growth is a common physiologic pattern characterized by short stature, slow but normal growth rate, and bone age of >2 SD (usually >2 years) behind the norm for chronological age. Because of the delayed maturation, pubertal development also lags behind and there is a late appearance of secondary sexual characteristics, although at the appropriate biological age as expressed by the bone age. Thus, height potential (again, based on genetic endowment) is generally preserved, although the more delayed an individual is, the less likely it is that he or she will meet the height prediction based on parental stature.[6] This pattern is called "constitutional delay of growth and adolescence", but it may also be due to mild to moderate systemic illness.

E. ATTENUATED GROWTH

Attenuated growth is characterized by growth failure, i.e., a subnormal growth velocity. Severe systemic or emotional illness is often a cause, but hormonal alterations — hypothyroidism, growth hormone (GH) deficiency, excessive glucocorticoids (endogenous or exogenous) and hypogonadism in adolescents — also will cause this pattern.

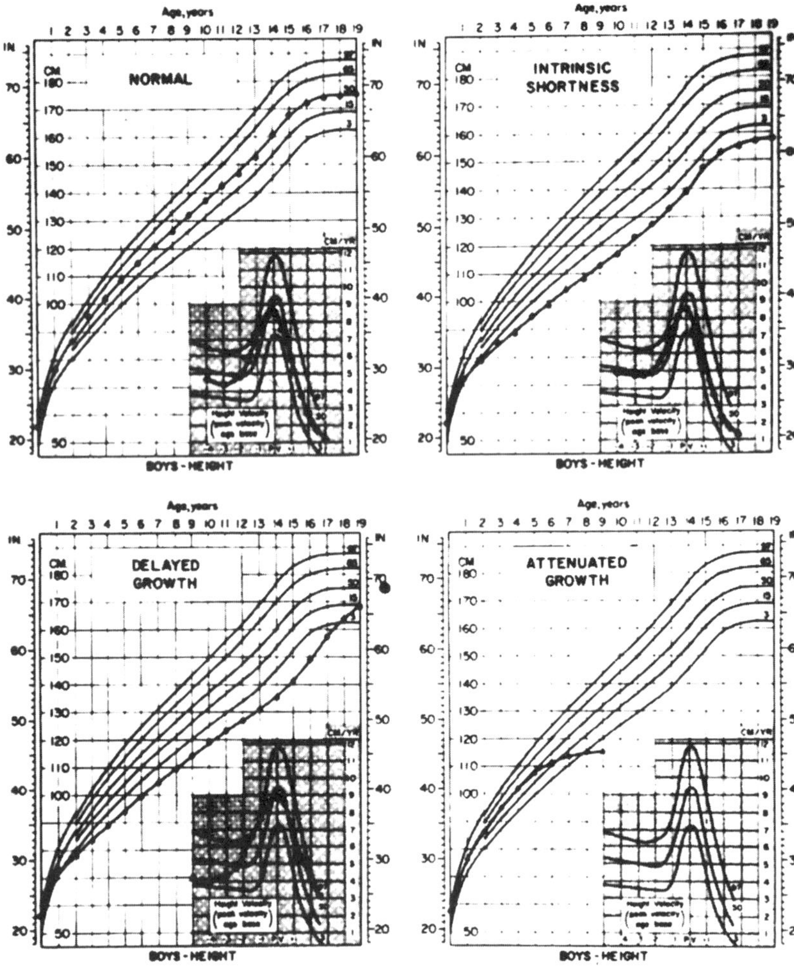

FIGURE 1. Linear growth curves in children with various types of growth disorders. Note that three prepubertal children of similar height at 9 years of age have different growth potential, depending on growth pattern. Growth curve in a child of average size is shown for comparison. (Insets) Height velocity charts based on years before or after peak pubertal height velocity. Normal percentiles are those of Tanner. (From Schaff-Blass, E. et al., *J. Pediatr.*, 104, 801, 1984. With permission.)

It is easy to recognize each of these patterns in isolation; however, it is common to have a combination because the physiological variants (intrinsic short stature and delayed growth) are so common. Anyone with growth failure deserves a full evaluation for systemic, emotional, and hormonal disorders, permitting a rational plan for therapy.

II. GROWTH HORMONE SECRETORY DYNAMICS

A. DEVELOPMENTAL ASPECTS

At all ages, fetal through adult, GH is secreted in an intermittent, pulsatile pattern. During childhood no differences in GH secretion exist between boys and girls, although several investigators have noted a significant positive correlation between physical stature and circulating levels for GH or between the amount of GH secreted per day and the height of the children.[7,8] In addition, Hindmarsh and colleagues[9] reported a relationship between height velocity and mean 24 h GH levels in short, prepubertal children. When this issue was investigated in more detail in short boys by Kerrigan and colleagues, no significant differences in pulsatile GH release were found between normally growing and short, prepubertal boys; however, a subset of short prepubertal boys with delayed bone age had subnormal GH release as indicated by low sums of GH pulse areas and of GH pulse amplitudes.[10–12] The finding of significant correlation in all subjects between growth velocity and the sum of GH pulse amplitudes is important because the results are compatible with the hypothesis that alterations of amplitude-modulated GH release underlie the pathophysiology of suboptimal growth in some short, prepubertal children.

B. DECONVOLUTION ANALYSIS

Deconvolution analysis was used to determine if the above abnormalities in GH secretion or metabolic clearance underlie the observed alterations in circulating hormone concentrations. Kerrigan and co-workers[10] found different secretory mechanisms were responsible for the indistinguishable mean overnight GH production rates in the normal vs. short vs. short-delayed bone age children. Both the mass of GH released per secretory burst and the maximal rate of hormone release were less for the short-delayed bone age boys compared to the normal and short boys without delayed bone age. Given the association of GH pulse amplitude with normal childhood growth, these results suggest a specific neuroendocrine mechanism that underlies suboptimal prepubertal growth in a subset of short, prepubertal children.

Major alterations in GH secretion occur at puberty in boys and girls.[13,14] Sex steroid hormones exert a major effect by augmenting GH pulse amplitude with little effect on pulse frequency. In late pubertal males the physiological mechanism contributing to the increased GH production rate is an augmentation of maximal rate of GH release per secretory episode. This results in a greater mass of GH released per secretory burst. No significant change occurred in the duration of each burst, the burst frequency, or serum GH half-life. In a similar manner, GH binding protein and body mass index (BMI = mass/height^2) vary throughout puberty and are both inversely related to GH secretion.[15,16]

The complex system in the general circulation to regulate the amount and pattern of GH secretion — GH, GH binding proteins, insulin-like growth factor-I (IGF-I) and IGF binding proteins and those derivatives of body composition that regulate them — precludes a simple relationship between circulating mean GH levels and linear growth velocity. Thus, it may be difficult in individual subjects to predict growth velocity or GH sufficiency using mean GH levels or the circulating GH levels. Evaluation of the mode of GH release as noted above remains a research tool, presently without practical therapeutic application to large groups of short, slowly growing children.

III. GROWTH HORMONE DEFICIENCY

A. DEFINITION

Growth hormone has not been considered necessary for normal growth during fetal life and in early infancy, but recent studies found that congenitally GH-deficient neonates have lower mean growth rates and are relatively overly fat for birth weight.[17,18] Most infants with GH deficiency show an abnormally decelerating growth velocity during the first 6 to 12 months of life and are >2 SD below the mean length for age by the end of the first year.[17,18]

Because no absolutely reliable tests for GH reserve (i.e., to diagnose GH deficiency) are available, several approaches were taken to determine the amount and pattern of GH secretion necessary for normal growth. Physiological stimuli (sleep and exercise); pharmacological stimulation (insulin-induced hypoglycemia, infusion of arginine, and the administration of L-DOPA or other secretagogues, orally); and spontaneous secretion (serial blood sampling at intervals of 10 to 20 min) were used to define GH sufficiency. It is beyond the scope of this chapter to discuss the pros and cons of each method. This subject has been evaluated in depth, but a general consensus remains elusive.[19,20] A single basal level of IGF-I is probably a useful measure, if normal; however, as nutritional factors play an important role in the regulation of IGF-I release, one may have a slightly low level of IGF-I, but still have normal GH release. The principal binding protein for IGF-I is IGF binding protein-3 (IGFBP-3), which is dependent on GH secretion. Thus, if the level of IGF-I is normal or nearly normal and that of IGFBP-3 is normal, GH deficiency is unlikely because IGF-I and IGFBP-3 reflect spontaneous GH secretion.[21]

Growth hormone deficiency is seldom absolute unless one is missing the structural gene for human GH (hGH). The diagnosis may be complicated by a constellation of physical and hormonal findings that are along a spectrum from low normal GH sufficiency to absent GH secretion. The laboratory evaluation of GH status must be in the correct biological context, i.e., in a child who is usually short (which may not be the case with recently acquired GH deficiency), who must have growth failure, a delayed bone age, and usually immature body proportions. The magnetic resonance imaging scan can be very helpful to confirm the diagnosis of idiopathic as well as organic causes of GH deficiency. A large percentage of the subjects with hypopituitarism have an ectopic neurohypophysis and absence of the infundibulum, likely indicating absence (interruption) of the portal vascular system.[22]

B. THERAPY

Although one cannot predict with certainty the response of a child to GH therapy, six variables predicted 40% of the variability in response to treatment in children with idiopathic GH deficiency (listed in relative importance: age, log maximal GH, weight adjusted for height, dosing schedule, dose, and mid-parental height). With the use of recombinant GH (natural sequence or with an extra NH_2 terminal methionyl residue) one can expect the growth rate to accelerate from 2 to 4 cm/year before treatment, to 10 to 15 cm/year for the first year, and 8 to 12 cm/year during the subsequent years. Pulsatile delivery of GH or GH releasing hormone (GHRH) to rodents permits greater growth than with the same total amount of GH injected once daily. Pulsatile delivery also causes greater increases of IGF-I messenger RNA levels within the epiphyseal growth plate. Intermittent delivery of GHRH and GH is effective in children, although not enough dose-response data exist to discern a differential effect between single and multiple daily administration of either compound. There are very few side effects — hyperinsulinism and mild glucose intolerance in this group of children and adolescents — and very little controversy about its efficacy.

Growth hormone therapy is associated with a redistribution of adipose tissue stores to more peripheral sites.[23] With GH treatment not only does a decline in fat mass and increase in lean body mass occur, but also a direct or indirect increase in the lipolytic rate and a decrease in the rate of re-esterification of the liberated free fatty acids.[24] Thus, GH affects many aspects of adipose tissue metabolism, and its exact role in the normal physiology (and pathophysiology of adipose tissue metabolism, as in obesity) remains unclear, especially concerning puberty, when these processes undergo the greatest change since fetal life.

C. TURNER SYNDROME

Turner syndrome occurs in girls who lack, or have a structural abnormality of, one of the X chromosomes. The syndrome is associated with short stature (approximately 20 cm deficit), gonadal dysgenesis, and multiple somatic abnormalities such as webbed neck, low set posterior hairline, high-arched palate, increased carrying angle at the elbow, and renal and cardiac anomalies. Although nearly all patients are short and slowly growing, a defined hormonal deficiency is not apparent.

Growth hormone secretion as well as IGF-I levels are normal prepubertally, but because of estrogen deficiency, do not increase at the expected time of adolescent development. A combination of the genetic abnormality (loss of genes affecting growth), skeletal malformations, and disordered lymphatic system, rather than an endocrine abnormality, is considered responsible for the short stature associated with Turner syndrome.

Although GH deficiency is not considered the pathophysiological mechanism responsible for short stature in girls with the Turner syndrome, GH therapy was initiated by a number of investigators. The results of most of the clinical trials are concordant. A marked rise in growth rate follows the start of GH therapy. Studies

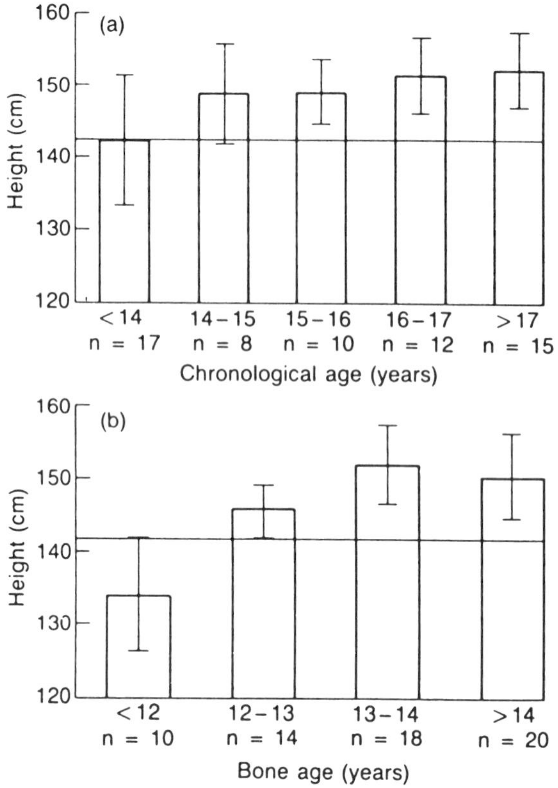

FIGURE 2. Comparison of (a) current height and chronological age and (b) current height and bone age of patients with Turner's syndrome treated for 4 to 7 years. Values are mean ± SD, and the horizontal lines represent the mean of the reference population described by Lyon et al. (From Rosenfeld et al., *Acta Pediatr. Scand.*, Suppl. 383, 1–6, 1992. With permission.)

with a long-term follow-up describe an augmentation in adult height when compared to historical control data of hundreds of girls with Turner syndrome. Growth acceleration can be sustained for at least 6 years. Although logical, the addition of low doses of estrogen to GH therapy does not appear to increase adult stature. Recently, the data have been disquieting showing that even the low doses of estrogen used are not low enough to prevent accelerated epiphyseal closure and an abrogation of the gains affected by GH alone. The addition of androgen (usually in the form of oxandrolone, a nonaromatizable, synthetic anabolic steroid) has been associated with impressive increases in growth velocity. Also, the current height of subjects 14 years of age or more, treated for 3 to 7 years with GH (some received oxandrolone as well), averaged >8 cm above the pretreatment projected final heights (Figures 2 and 3).[25] These results probably represent the lower boundary of the potential increase in adult height because many girls began to receive growth-promoting therapy relatively late. In addition to the increase in

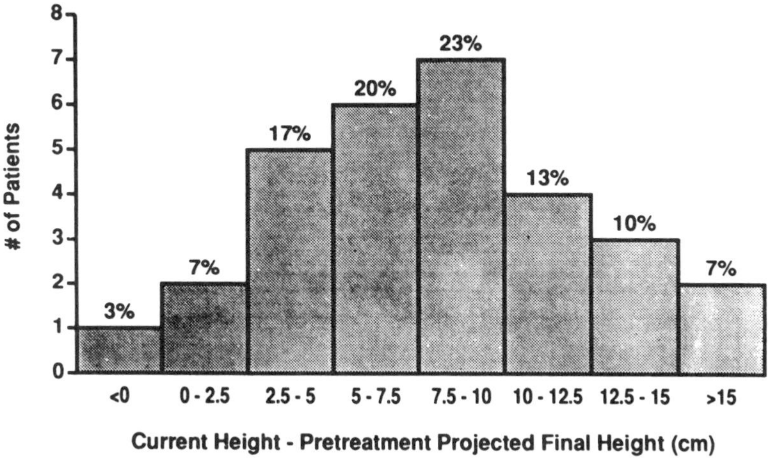

FIGURE 3. Current height minus pretreatment projected final height[3] for the 30 subjects who terminated GH treatment. Numbers above each bar indicate percentage of group of 30 subjects. (From Rosenfeld, R. G. et al., *J. Pediatr.*, 121, 49, 1992. With permission.)

final height, one must consider the psychosocial benefits of normalization of height during childhood and adolescence.

D. RENAL FAILURE

Children with chronic renal insufficiency (CRI) or end-stage renal disease (ESRD) do not reach their genetic height potential and often have an adult height >2 SD below the norm despite delayed adolescent development. The GH axis is likely disturbed with increased GH and decreased IGF-I bioactivity. Because of the abnormal clearance of IGFBPs (especially IGFBP-1 and IGFBP-3 and its components), the levels of IGF-I may in fact be elevated, but the biological activity is depressed because of increased attachment to the binding protein. The biologically effective IGF-I is thus sequestered in the plasma component and rendered physiologically ineffective.

Neither GH deficiency nor thyroid hormone insufficiency appear responsible for the slow growth in CRI or ESRD. Spontaneous GH release and clearance, however, are affected by CRI. Deconvolution analysis showed an increased frequency of secretory events, but with lessened amplitude (maximal secretory rate). The clearance is markedly prolonged.[26] These defects appear due to decreased feedback inhibition of GH secretion by pituitary/hypothalamic IGF-I bioactivity. The abnormality in the GH-IGF-I axis may not reflect CRI per se, but merely be a marker for starvation. Inadequate energy intake, especially during infancy, with its higher requirement per kilogram, may lead to poor growth in CRI. Children with the lowest caloric intakes grow the least. Other possible causes for slow growth include circulating uremic toxins and inhibitors (generally considered to be IGFBPs, especially the small molecular weight forms (20 to 30 kDa) of IGFBP-3 that can escape the vascular compartment), renal osteodystrophy with its attendant secondary hyperparathyroidism, and prednisone or other drugs. Most patients reaching ESRD before and during puberty irreversibly lose growth potential during puberty. Their growth spurt is most often delayed and is less than the normal maximum.

E. TREATMENT

Because neither dialysis nor subsequent successful renal transplantation restores the genetic height potential nor uniformly leads to growth acceleration (''catch-up'' growth), a number of therapeutic trials with growth hormone were undertaken.

Short-term studies showed that administration of hGH can increase the growth rate in children with CRI managed with or without dialysis. Growth hormone was much more effective than placebo over a 6-month trial.[27] Levels of IGF-I and IGFBP-3 (not its fragments) increased, while levels of IGFBP-1, thought to diminish IGF-I bioactivity, decreased. Further clinical trials indicated efficacy of GH as measured by continued acceleration of growth for at least 3 years.[28] As the bone age did not accelerate to a similar degree, optimism is high for improved adult height. Preliminary data of growth acceleration with GH therapy following renal transplantation were presented by van Dop and colleagues.[29] These patients are often complicated by the large number of drugs received, especially glucocorticoids.

In summary, the causes of short stature and diminished growth rate in patients with CRI, ESRD, and following renal transplantation are complex and include endocrine, but also many non-endocrine, causes. Growth hormone administration holds promise for accelerating the growth rate and even increasing adult height; however, some (disquieting) evidence in children suggests that treatment with GH may adversely affect renal function and may even accelerate the rate of loss of nephron function.[30] Given the multiple dysfunctions in chronic renal failure, it may make sense to combine GH, calcitonin, and erythropoietin with aggressive nutritional therapy to maximally accelerate growth in prepubertal, but especially in pubertal, children. If renal transplantation is undertaken, it is imperative to use the least amount of glucocorticoid possible to maintain adequate immunosuppression, but permit growth. Similar to other chronic conditions for which glucocorticoids are used, an every other day regimen is preferable to maximize the growth rate.[31]

F. NONGROWTH HORMONE DEFICIENCY

A large number of trials of GH therapy were undertaken to augment the growth velocity in short, slowly growing children. Most protocols included otherwise normal prepubertal children whose birth weights were >2.5 kg (i.e., not intrauterine growth retarded) and whose responses to GH provocative tests were within the normal range. The largest group of children treated under these protocols

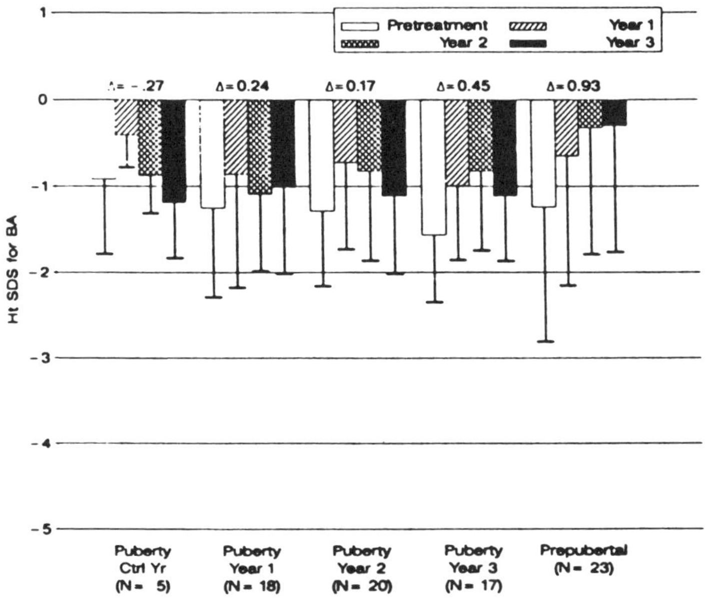

FIGURE 4. Bayley-Pinneau PAH SDS for each treatment year by pubertal group (mean ± SD). The change in Bayley-Pinneau PAH (Δ) from before treatment to year 3 of treatment is shown for the pubertal groups. Although overall gains were significant in each group ($p < 0.05$), no statistical differences were found among groups. (From Hopwood, N. J. et al., *J. Pediatr.*, 123, 215, 1993. With permission.)

represents variants of normal: constitutional delay of growth, familial short stature, and a combination of these two common variants, producing the shortest and slowest growing "normal" children. Preliminary studies of this heterogeneous group of subjects suggest that a majority will significantly increase their growth rate in response to GH treatment.[32] A larger study with a nontreated control group showed a statistically significant increase in the growth rate for GH-treated, prepubertal children with an increase in relative height for age — from -2.7 ± 0.5 to -2.2 ± 0.6 SD; $p < 0.001$ — while the control group remained at -2.8 SD.[12] Although promising, these data are too preliminary to indicate that increased adult height is likely, as was already shown for Turner syndrome girls (see above). The longitudinal data were evaluated after 3 years.[33] Year 1 growth velocity expressed as change in height SD score was the best predictor of subsequent growth response. However, responses to therapy could not be reliably predicted from baseline growth or hormonal data (Figure 4).

Bierich and colleagues treated a relatively homogeneous group of constitutionally delayed boys with GH in a dose of 6 to 8 mg/m^2 for an average of 3 years (range 2.5 to 6 years).[34] At the end of therapy the boys had moved from -3.2 to -2.4 SD (height for age), and then without additional therapy to -1.6 SD as adults (height was measured). The final height corresponded exactly to the height predicted before therapy. Thus, these data are consistent with an accelerated tempo of growth during GH therapy, but with permanent short stature of moderate degree and relative to parents' stature.

Intrauterine growth retardation also represents a heterogenous group of short, slowly growing children. Several small studies were undertaken with GH therapy in children with this diagnosis. One longitudinal study of children with birth weights below the third percentile adjusted for gestational age, who received GH for 3 years, found mean height velocity SD score to be +1.1 at the end of the trial.[35] However, the bone age also accelerated slightly so that no evidence (height for bone age SD score) was found for increased final height potential. Thus, in this heterogeneous group of patients with variable growth responses to exogenous hGH, only an alteration in the tempo of growth appeared, as was seen in children with delayed maturation.

REFERENCES

1. **Lacey, K. A. and Parker, J. M.,** The normal short child: community study of children in New Castle-upon-Tyne, *Arch. Dis. Child.,* 49, 417, 1974.
2. **Horner, J. M., Thorsen, A. V., and Hintz, R. L.,** Growth deceleration patterns in children with constitutional short stature: an aid to diagnosis, *Pediatrics,* 62, 529, 1978.
3. **Greulich, W. W. and Pyle, S. I.,** *Radiographic Atlas of Skeletal Development of the Hand and Wrist, 2nd ed.,* Stanford University Press, Stanford, CA, 1959.
4. **Smith, D. W., Truog, W., Rogers, J. E. et al.,** Shifting linear growth during infancy: illustration of genetic factors in growth from fetal life through infancy, *J. Pediatr.,* 89, 225, 1976.
5. **Schaff-Blass, E., Burstein, S., and Rosenfield, R. L.,** Advances in the diagnosis and treatment of short stature with special reference to the role of growth hormone, *J. Pediatr.,* 104, 801, 1984.
6. **Blethen, S. L., Gaines, S., and Weldon, V.,** Comparison of predicted and adult heights in short boys: effect of androgen therapy, *Pediatr. Res.,* 104, 182, 1984.
7. **Albertsson-Wikland, K. and Rosberg, S.,** Analyses of 24-hour growth hormone profiles in children: relation to growth, *J. Clin. Endocrinol. Metab.,* 67, 493, 1988.
8. **Albertsson-Wikland, K., Rosberg, S., Libre, E., Lundberg, L.-O., and Groth, T.,** Growth hormone secretory rates in children as estimated by deconvolution analysis of 24-h plasma concentration profiles, *Am. J. Physiol.,* 257, E809, 1989.
9. **Hindmarsh, P., Smith, P. J., Brook, C. G. D., and Matthews, D. R.,** The relationship between height velocity and growth hormone secretion in short prepubertal children, *Clin. Endocrinol.,* 27, 581, 1987.
10. **Kerrigan, J. R., Martha, P. M., Jr., Blizzard, R. M., Christie, C. M., and Rogol, A. D.,** Variations of pulsatile growth hormone release in healthy short prepubertal boys, *Pediatr. Res.,* 28, 11, 1990.
11. **Kerrigan, J. R., Martha, P. M., Jr., Veldhuis, J. D., Blizzard, R. M., and Rogol, A. D.,** Altered growth hormone secretory dynamics in prepubertal males with constitutional delay of growth, *Pediatr. Res.,* 33, 278, 1993.
12. **Veldhuis, J. D., Blizzard, R. M., Rogol, A. D., Martha, P. M., Jr., Kirkland, J. L., Sherman, B. M., and Genentech Collaborative Group,** Properties of spontaneous growth hormone secretory bursts and half-life of endogenous growth hormone in boys with idiopathic short stature, *J. Clin. Endocrinol. Metab.,* 74, 766, 1992.

13. **Kerrigan, J. R. and Rogol, A. D.,** The impact of gonadal steroid hormone action on growth hormone secretion during childhood and adolescence, *Endocr. Rev.*, 13, 281, 1992.
14. **Rose, S. R., Municchi, G., Barnes, K. M., Kamp, G. A., Uriarte, M. M., Ross, J. L., Cassorla, F., Cutler, G. B., Jr.,** Spontaneous growth hormone secretion increases during puberty in normal girls and boys, *J. Clin. Endocrinol. Metab.*, 73, 428, 1991.
15. **Martha, P. M., Jr., Rogol, A. D., Blizzard, R. M., Shaw, M. A., and Baumann, G.,** Growth hormone-binding protein activity is inversely related to 24 hour growth hormone release in healthy boys, *J. Clin. Endocrinol. Metab.*, 73, 175, 1991.
16. **Martha, P. M., Jr., Gorman, K. M., Blizzard, R. M., Rogol, A. D., and Veldhuis, J. D.,** Endogenous growth hormone secretion and clearance rates in normal boys, as determined by deconvolution analysis: relationship to age, pubertal status, and body mass, *J. Clin. Endocrinol. Metab.*, 74, 336, 1992.
17. **Gluckman, P. D., Gunn, A. J., Wray, A., Cutfield, W. S., Chatelain, P. G., Guilbaud, O., Ambler, G. R., Wilton, P., and Kerstin, A.-W.,** Congenital idiopathic growth hormone deficiency associated with pre-natal and early post-natal growth failure, *J. Pediatr.*, 121, 920, 1992.
18. **Wit, J. M. and van Unen,** Growth of infants with neonatal growth hormone deficiency, *Arch. Dis. Child.*, 67, 920, 1992.
19. **Rose, S. R., Ross, J. L., Uriarte, M., Barnes, K. M., Cassorla, F. G., and Cutler, G. B., Jr.,** The advantage of measuring stimulated compared with spontaneous growth hormone levels in the diagnosis of growth hormone deficiency, *N. Engl. J. Med.*, 319, 201, 1988.
20. **Bercu, B. B., Shulman, D., Root, A. W., and Spiliotis, B. E.,** Growth hormone (GH) provocative testing frequently does not reflect endogenous GH secretion, *J. Clin. Endocrinol. Metab.*, 63, 709, 1986.
21. **Blum, W. F., Albertsson-Wickland, K., Rosberg, S., and Ranke, M. B.,** Serum levels of insulin-like growth factor I (IGF-I) and IGF binding protein 3 reflect spontaneous growth hormone secretion, *J. Clin. Endocrinol. Metab.*, 76, 1610, 1993.
22. **Abrahams, J. J., Trefelner, E., and Boulware, S.D.,** Idiopathic growth hormone deficiency: MR findings in 35 patients, *Am. J. Neuroradiol.*, 12, 155, 1991.
23. **Rosenbaum, M., Gertner, J. M., Gidfar, N., Hirsch, J., and Leibel, R. L.,** Effects of systemic growth hormone (GH) administration regional adipose tissue in children with non-GH-deficient short stature, *J. Clin. Endocrinol. Metab.*, 75, 151, 1992.
24. **Quabbe, H. J., Bratzke, H. J., Siegers, U., and Elban, K.,** Studies on the relationship between plasma free fatty acids and growth hormone secretion in man, *J. Clin. Invest.*, 5, 2388, 1972.
25. **Rosenfeld, R. G., Frane, J., Attie, K. M., Brasel, J. A., Burstein, S., Cara, J. F., Chernausek, S., Gotlin, R. W., Kuntze, J., Lippe, B. M., Mahoney, P. C., Moore, W. V., Saenger, P., and Johnson, A. J.,** Six-year results of a randomized, prospective trial of human growth hormone and oxandrolone in Turner syndrome, *J. Pediatr.*, 121, 49, 1992.
26. **Veldhuis, J. D., Johnson, M. L., and Bolton, W. K.,** Analysis of pulsatile endocrine data in patients with chronic renal insufficiency: a brief review of deconvolution techniques, *Pediatr. Nephrol.*, 5, 522, 1991.
27. **Hokken-Koelega, A. C. S., Stijnen, T., de Muinck Keizer-Schrama, S. M. P. F., Wit, J. M., Wolff, E. D., de Jong, J. C. J. W., Donckerolcke, R. A., Abbad, N. C. B., Bot, A., Blum, W. F., and Drop, S. L. S.,** Placebo-controlled double-blind, cross-over trial of growth hormone treatment in prepubertal children with chronic renal failure, *Lancet*, 338, 585, 1991.
28. **Fine, R. N.,** Stimulating growth in uremic children, *Kidney Int.*, 42, 188, 1992.
29. **Van Dop, C., Jabs, K. L., Donohoue, P. A., Bock, G. H., Fivush, B. A., and Harmon, W. E.,** Accelerated growth rates in children treated with growth hormone after renal transplantation, *J. Pediatr.*, 120, 244, 1992.
30. **Andersson, H. C., Markello, T., Schneider, J. A., and Gahl, W. A.,** Effect of growth hormone treatment on serum creatinine concentration in patients with cystinosis and chronic renal disease, *J. Pediatr.*, 120, 716, 1992.

31. **Broyer, M., Guest, G., and Gagnadoux, M.-F.,** Growth rate in children receiving alternate-day cortico steroid treatment after kidney transplantation, *J. Pediatr.*, 120, 721, 1992.
32. **Van Vliet, G., Styne, D. M., Kaplan, S. L., and Grumbach, M. M.,** Growth hormone treatment for short stature, *N. Engl. J. Med.*, 309, 1016, 1983.
33. **Hopwood, N. J., Hintz, R. L., Gertner, J. M., Attic, K. M., Johanson, A. J., Baptista, J., Kuntze, J., Blizzard, R. M., Cara, J. F., Chernausek, S. D., Kaplan, S. L., Lippe, B. M., Plotnick, L. P., and Saenger, P.,** Growth response of children with non-growth-hormone deficiency and marked short stature during three years of growth hormone therapy, *J. Pediatr.*, 123, 215, 1993.
34. **Bierich, J. R., Nolte, K., Drews, K., and Brügmann, G.,** Constitutional delay of growth and adolescence. Results of short-term and long-term treatment with GH, *Acta Endocrinol.*, 127, 392, 1992.
35. **Stanhope, R., Preece, M. A., and Hamill, G.,** Does growth hormone treatment improve final height attainment of children with intrauterine growth retardation?, *Arch. Dis. Child.*, 66, 1180, 1991.

Chapter 7

GROWTH HORMONE RECEPTOR DEFICIENCY

Mary A. Vaccarello

TABLE OF CONTENTS

I. INTRODUCTION

Over a quarter of a century ago, Zvi Laron and colleagues described three Israeli siblings with the phenotypic features of growth hormone (GH) deficiency but with elevated levels of immunoreactive GH.[1] It is now known that the clinical and biochemical phenotype that characterizes this condition is due to inadequate synthesis of insulin-like growth factor-I (IGF-I), the endocrine hormone responsible for promoting end-organ growth in response to GH.

Our understanding of the pathogenesis of Laron syndrome (LS) reflects research efforts that have also contributed to our current knowledge of the regulatory factors that influence the GH-IGF-IGFBP (BP, binding protein) axis and autocrine/paracrine function. It is now generally accepted that LS constitutes the primary form of GH insensitivity syndrome (GHIS), and that growth hormone receptor deficiency (GHRD), a subset of LS, results from a qualitative or quantitative alteration of the GH receptor.[2] This chapter reviews the clinical, biochemical, and genetic characteristics of LS, with special emphasis on GHRD, and focus on the therapeutic potential of recombinant IGF-I.

0-8493-8384-6/95/$0.00+$.50

TABLE 1
Growth Hormone Insensitivity Syndrome (Classification)[2]

Primary GH insensitivity syndromes (Laron syndrome)
GH receptor deficiency (quantitative or qualitative defects of the GH receptor)
Postreceptor defects (abnormalities of GH receptor signal transduction)
Primary defect of IGF-I synthesis
Secondary causes of GH insensitivity syndrome (acquired or transitory)
Antibodies to GH that inhibit GH action
Antibodies to the GH receptor
Malnutrition
Liver disease
Other

A. TERMINOLOGY AND CLASSIFICATION

The current accepted terminology of this condition, "Laron syndrome", was introduced in an attempt to avoid the pejorative connotation of "dwarfism" which serves as a social obstacle for many affected individuals.[3]

Growth hormone resistance is a characteristic finding in LS but is also found in many other conditions. Thus, the term GHIS was introduced to describe conditions associated with GH resistance. In 1992 at the First International Workshop on Growth Hormone Insensitivity held in Amsterdam, researchers met to develop a classification for GHIS.[2] The syndrome was divided into two subtypes: primary (Laron syndrome), related to receptor/postreceptor defects or altered IGF-I synthesis, and secondary, associated with a primary condition or disease (Table 1). Patients with the clinical and biochemical features of LS are considered to represent the "pure" form of GHIS regardless of where the defect lies.

B. EPIDEMIOLOGICAL FACTORS

Laron syndrome occurs predominantly in people of Mediterranean and Middle Eastern origins, and in families with a high frequency of consanguinity.[1,3,4] The two largest populations reported thus far, from Israel and southern Ecuador, demonstrate these epidemiological characteristics well. All Ecuadorian patients are descendants of Spanish ancestors and originate from two provinces, Loja and El Oro. Common family names are recognized to be of those who converted from Judaism to Catholicism during the Inquisition (*conversos*) and historic evidence exists of their ancestors' early 16th century immigration to Ecuador.

II. HISTORICAL REVIEW

The historical events leading to the characterization of LS began in 1966 with the original description of three Jewish siblings of Yemenite origin.[1] By 1968, reports appeared of other affected individuals from Oriental Jewish Israeli families with a high degree of consanguinity.[4] The pattern of inheritance demonstrated within these kindreds was consistent with that of an autosomal recessive gene with full penetrance. The initial assumption that the GH molecule of these individuals was biologically inactive was based on the reported response to exogenous GH, i.e., increased plasma free fatty acids, hypercalciuria, and retention of nitrogen, water, and sodium.[1] In 1968, however, Merimee et al.[5] reported the failure of GH to produce an anabolic effect in one patient and hypothesized that the defect could be specific to the GH receptor. One year later, Daughaday et al.[6] found an absence of sulfation factor activation by GH in the plasma of six affected children, data which supported the GH receptor hypothesis. The GH effects initially noted may have been in response to hormonal contaminants of pituitary GH. Studies done during the 1970s confirmed that the GH molecule was unaltered in LS in terms of fractional distribution on Sephadex column, immunoreactivity, and binding affinity to donor liver membranes.[7–9]

Definitive evidence of a GH receptor defect was presented in 1984 when Eshet et al.[10] reported the failure of liver microsomes from two LS patients to bind radiolabeled GH. In 1986 IGF-I was synthesized by recombinant DNA technology making IGF-I replacement a therapeutic alternative for patients with LS.[11] Geffner et al.[12] demonstrated a clonogenic response to IGF-I but not to GH in T cells from patients with LS, suggesting that exogenous IGF-I could bypass a GH receptor defect. In hypophysectomized rodents recombinant IGF-I enhanced growth and body weight in the absence of GH.[13] The first trial involving IGF-I administration to healthy adults was reported in 1987 and prompted the first short-term trial in patients with LS the following year.[14,15] Growth hormone binding protein (GHBP) was reportedly undetectable in sera of patients with LS and presumed to reflect GH receptor function, as normal levels were detected in sera of GH-deficient children.[16] The subsequent purification, cloning, and sequencing of the GH receptor confirmed that the naturally occurring GHBP was structurally identical to the extracellular domain of the GH receptor and was postulated to result from proteolytic cleavage.[17]

Some 80 cases of LS were reported worldwide by 1989, with about half of the patients being of Oriental Jewish or Arab heritage. During this year the first mutations of the GH receptor gene were described in patients with LS and included deletions and point mutations specific to exons 3, 5, and 6.[18,19] It became apparent that this clinically homogeneous condition was the result of a heterogeneous genetic defect of the GH receptor.

In 1990 and 1991 Rosenbloom et al.[20] and Guevara-Aguirre et al.[3] described the largest concentration to date of GHRD from two adjacent provinces in southern Ecuador. The description of these Ecuadorian patients, believed to be inbred descendants of *conversos,* raised the possibility of a relationship with the Oriental Jewish families described 24 years before. That same year, several LS patients without a known defect of the GH receptor were reported to have normal GHBP levels, suggesting that a cytoplasmic or postreceptor defect may also be implicated in the pathogenesis of this disease.[21] The relationship between GHBP levels and receptor number or binding affinity was questioned when an Ecuadorian patient with a proven genetic defect of the extracellular domain of the GH receptor was also found to have low-normal levels of GHBP.[22]

The collaborative efforts of Cunningham et al.[23] and De Vos et al.[24] led to the discovery that two GH receptor molecules are necessary for the binding of one molecule of GH. They postulated that dimerization, the interaction of the carboxy terminal ends of the extracellular domain of the GH receptor, may stabilize binding of the second receptor and be necessary for signal transduction. A defect in dimerization would presumably present as LS/GHRD.

In 1992 Berg et al.[25] identified a point mutation of codon 180 in 36/37 Ecuadorian patients with GHRD, demonstrating the highest degree of genetic homogeneity within a GHRD population to date. The first reports of enhanced growth velocity in GHRD patients during long-term IGF-I administration were encouraging and subsequent reports confirmed that the growth response is significant.[26–31]

The first non-Ecuadorian patient found to be homozygous for the codon 180 point mutation was reported in 1993, an Israeli Jew of Moroccan parents.[32] The hypothesis that GHRD may have been carried to Ecuador by a settler of Jewish descent during the time of the Inquisition is supported by this observation.

To date, 197 cases have been reported worldwide. Genetic heterogeneity, variable phenotypic expression, and clustering in Mediterranean and Ecuadorian populations is common. The Ecuadorian cohort is the largest single population in the world afflicted with GHRD and now numbers 64.[33] The Ecuadorian patients are, with a single exception, genetically homogeneous (55/56) while exhibiting considerable clinical and biochemical variability.[25,33]

III. CLINICAL PHENOTYPE

The clinical appearance of individuals with GHRD is indistinguishable from GHD type 1A, which is due to a deletion of the GH gene.[34,35] The most striking features of GHRD are severe growth failure and immature facial development, making some infants recognizable at birth. Although birth weight is generally normal, 3 of 34 Ecuadorian children had birth weights <3 SD of the mean for North American newborns and Laron reported 12 of 20 newborns whose lengths were <2 SD of the mean for sex and ethnic origin.[33,34] Taken together these data suggest only mild intrauterine growth retardation in 27% of cases. Ecuadorian children with GHRD fell well below the normal growth curves for age and family. Mean parental and unaffected sibling heights ranged from 0 to -4 SD below the mean for North Americans, while heights of affected children ranged from -6.8 to -9.6 SD, and adults from -5.3 to -11.5 SD.[22,33] Final heights among the two populations differ in that the mean for 24 Ecuadorian females was 111.5 ± 7.5 compared to 119 ± 8.5 for 14 Israeli females, and 118 ± 11.7 ($n = 11$) compared to 124 ± 8.5 ($n = 10$) in males, respectively.

No significant effect of heterozygosity on stature was demonstrated among the Ecuadorian population. The mean height of 41 GHRD relatives who tested positive for heterozygosity was -1.81 ± 1.15 SD, while the mean height for the 24 siblings who are homozygous normal was -1.31 ± 0.95 ($p = 0.08$).[33]

Craniofacial abnormalities are most exaggerated during childhood and improve with growth. Although the head circumference is normal for height age, the foreshortened face and long forehead give the appearance of a large head.[36] The prominent forehead, shallow orbits, hypoplasia of the nasal bridge, and shortness of the face are characteristic. Of Ecuadorian children <10 years of age, 25% demonstrate the "setting sun" sign, sclera noted above the iris upon direct gaze, which may be a consequence of shallow orbits and prominent forehead combined.[33] In younger children the increased head circumference for height, delayed closure of the fontanelles, and prominence of cephalic veins may give the impression of hydrocephalus. Prolonged retention of primary teeth and dental decay is common in children, whereas adult dentition, although crowded, appears healthy. Thin, sparse hair growth and blue scleras have been reported in affected children from Ecuador and Israel.[22,37]

Seventy percent of Ecuadorian patients had a history of delayed walking attributed to hypomuscularity. High-pitched voices, common among children (100%) and adults (86%), may be associated with a small larynx.[33,36] Lean body mass is decreased relative to body fat, giving the appearance of obesity despite normal to decreased weight for height. Body weight for height increases with the onset of puberty, however, and ranges from 125 to 240% of ideal body weight in females and 110 to 170% in males. Osseous maturation is typically delayed in GH-deficient states; however, the bone age:height age ratio in GHRD (1.6 to 6.4) markedly exceeds that seen in isolated GH deficiency, with the exception of familial GH deficiency type 1A.[33,35]

Differences in body proportions between the GHRD child and adult may reflect the importance of GH/IGF-1 activity during puberty. Although upper:lower body segment ratios are normal for bone age in Ecuadorian GHRD children, 75% of adults have immature body proportions with upper : lower body segment ratios that are >1.1.[33] In contrast, in the Israeli population upper : lower segment ratios were increased when analyzed for chronological age.[33,38] Arm span for bone age is normal in children, while adults have an average reduced arm span for height of 8 cm.[33] The appearance of small hands and feet for stature are consistent findings in 70% of Ecuadorian patients, who had hands or feet that were <10th percentile for height.

Sexual development is also affected in GHRD. Male phallic length in children was described as small in 9 of 13 European boys and <2 SD from the mean for bone age in 10 of 10 Ecuadorian boys.[33,39] Full sexual maturation is achieved in males and although adult testicular size appears small, it is proportional to body size. Timing of puberty among Israeli patients has been described as normal in females and delayed in males, whereas 50% of Ecuadorian men and women experienced delays of up to 7 years.[33,40] Fertility has been well documented in males and females with GHRD.[40,41]

Disparity in mortality exists among Ecuadorian patients and their siblings.[3,22] Among GHRD children <7 years of age mortality was 19% compared to 11% among nonaffected siblings of the same age ($p < 0.05$).[3] Mortality was due to common childhood infectious diseases in both groups. No difference is apparent in mortality among adults; a 66-year-old female and a 55-year-old male died of apparent myocardial infarction and two sexagenarians remain in good health.

The two large population cohorts differ in intellectual status.[42,43] Israeli patients reportedly have subnormal intelligence and serious psychosocial problems.[34,43] Among affected Ecuadorians who attended school, 69% were considered to be of the highest academic standing and almost all were reported to be of above average intelligence when compared to family members.[42] These differences require further examination in a controlled study with intelligence measures.

IV. BIOCHEMICAL PHENOTYPE

The biochemical characteristics of LS result from the inability of GH to effectively bind to the hepatic receptor and induce IGF-I synthesis.[12] Low circulating levels of IGF-I are incapable of inducing a negative feedback effect on GH through the release of somatostatin, the inhibition of GHRH, or the direct suppression of somatotrophs.[44] This results in elevated levels of circulating GH in an individual with the clinical phenotype of GH deficiency.

A. GROWTH HORMONE

In the original description of LS elevated serum levels of GH were detected in three siblings (57 to 88 μg/l).[1] Enhanced responsiveness of serum GH levels to arginine and, paradoxically, following intravenous glucose, was subsequently demonstrated.[4] In the Ecuadorian GHRD population the mean nonstimulated GH level from 19 children was significantly increased (all >10 μg/l) when compared to the mean for affected adults, revealing an age-related association (Table 2).[3,33] The decline in serum GH levels after puberty is similar to that seen in normal adults and may reflect the stimulatory effect of sex hormones on IGF-I synthesis and subsequent feedback inhibition of IGF-I on GH secretion. Studies of GH secretion in LS have demonstrated pulsatile secretion, increased production, and normal metabolic clearance.[45] Evaluation of 24-h diurnal patterns for GH in six Ecuadorian adults with GHRD showed marked variability among subjects.[46] In three patients, elevated spontaneous GH secretion was noted, while normal secretory dynamics were observed in the remaining three. On analysis of the mean data significant suppression of GH release was observed with IGF-I therapy (40 μg/kg every 12 h); integrated 24-h GH level (6.5 ± 2.1 to 1 ± 0.2 μg/l), number of peaks (8.3 ± 0.6 to 5.7 ± 0.95), area under the curve (5413 ± 1732 to 592 ± 171 μg/l), peak height (11.9 ± 3.4 to 2.8 ± 0.5 μg/l), and peak clonidine-stimulated GH level (17 ± 3.7 to 5.8 ± 2.1 μg/l).

TABLE 2
Serum Biochemical Features of GHRD in Ecuador[22]

		Control (n = 22 adults)	Children ≤ 16 years (n = 19)	Adults ≥ 16.1 years (n = 31)
IGF-I	(μg/l)	96–270	3 ± 2	25 ± 19*
IGF-II	(μg/l)	388–792	70 ± 42	151 ± 76*
IGFBP-2	(% control)	100	290 ± 123	173 ± 111**
IGFBP-3	(μg/l)	2500 (pooled)	226 ± 173	433 ± 149**
GH	(μg/l)		32 ± 22	11 ± 11*

Note: IGFBP-2 was measured by Western ligand blotting. All other measurements were analyzed by radioimmunoassay, with the exception of some GH values which were obtained by immunoradiometric assay. Control data were obtained from healthy Ecuadorian volunteers. Significant differences for data obtained for children and adults are represented by an *(p <0.0001) and **(p <0.001).

B. INSULIN-LIKE GROWTH FACTORS

Sulfation factor activity, a measurement of IGF-I bioactivity, was initially assessed in LS by quantitating the uptake of radiolabeled sulfate by cartilage of hypophysectomized rats in the presence of varying dilutions of human sera.[6] Serum from LS patients was incapable of enhancing the uptake of radioactive sulfate, while GHD sera achieved a sustained increase only in the presence of GH. Among a heterogeneous European population of presumed GHRD subjects all subjects had serum levels of IGF-I that were <5th percentile for age and 25 of 27 had levels that were <0.1 percentile.[39] Insulin-like growth factor-II values were also <5th percentile for age (62 to 232 μg/l). In the Ecuadorian GHRD population serum IGF-I levels were markedly reduced and an age-related effect was identified (Table 2).[22] Increased IGF-I levels in adults, despite declining GH levels, suggest that sex hormones enhance IGF-I production independent of GH.

C. INSULIN-LIKE GROWTH FACTOR BINDING PROTEINS

There are currently six well-characterized binding proteins with high affinity for IGFs. These IGFBPs appear to modulate the actions of IGFs at the target tissue.[47] The GH-dependent IGFBP-3 serves as a reservoir for IGF-I in serum where it is detected as a 150-kDa protein complexed to an acid labile subunit (85 to 88 kDa) and either IGF-I or II (7.5 kDa). Hardouin et al.[48] initially characterized uncomplexed IGFBP-3 in the sera of LS patients by Western ligand blotting (WLB) and observed a characteristic doublet (38.5 to 41.5 kDa) that was less intense than that seen in patients with familial GH deficiency. In contrast, the 30 and 34 kDa bands were increased (IGFBP-1 and -2, respectively). Western ligand blotting analysis of sera from Ecuadorian patients revealed that IGFBP-3 levels in children ranged from 0 to 9% of normal and in adults from 4 to 58%, while IGFBP-1 and -2 were moderately increased in children and adults.[33] Evaluation of serum IGFBP-3 by radioimmunoassay in the Ecuadorian GHRD population revealed similar findings. Mean IGFBP-3 levels were markedly low when compared to pooled serum from Ecuadorian controls (Table 2).[22] The age-related association seen by WLB was confirmed and IGFBP-3 levels demonstrated an inverse correlation with statural deviation in children as well as in adults.[33]

It was assumed that IGF-I therapy would stimulate IGFBP-3 synthesis in humans because serum IGFBP-3 increased in hyophysectomized rodents after IGF-I administration.[49] The data obtained from IGF-I therapeutic trials, however, do not confirm this assumption.[46,50]

D. GROWTH HORMONE BINDING PROTEIN (GHBP)

GHBP is a single-chain glycoprotein with a molecular weight of 60 kDa which binds 1 mole of GH for each mole of binding protein and is highly species specific.[17,51] A marked reduction of GHBP was observed in GHRD sera evaluated by gel filtration of radiolabeled GH.[16] Similar results were obtained when GHBP was measured by chromatography in the sera of 20 Ecuadorian patients (1 to 30% of control).[20] Subsequently, Carlsson et al.[52] developed a ligand-mediated immunofunctional technique which employs an anti-GH monoclonal antibody and an anti-GH antibody conjugated to horseradish peroxidase for measurement of the functional capacity of GHBP to bind to GH. The binding capacity of GHBP in the sera of 49 Ecuadorians with GHRD was evaluated and compared to Ecuadorian controls (308 ± 184 pmol/l in females and 230 ± 118 pmol/l in males).[22] Even though all patients had the same GH receptor gene mutation, only 53% had undetectable levels of GHBP binding capacity (<31 pmol/l) and 8 patients had levels that were >40% of the low range for sex-specific controls. Growth hormone binding protein did not correlate with stature, serum IGF-I levels, or age. Assessment of 27 subjects with GHRD from Europe and Australia identified 18 patients with undetectable [^{125}I]hGH binding to GHBP, 2 with markedly reduced binding, and 7 females who demonstrated normal [^{125}I]hGH binding.[53] Subjects with LS and normal levels of GHBP activity are otherwise indistinguishable from patients with undetectable GHBP activity.

E. CAN HETEROZYGOSITY BE DETERMINED BY BIOCHEMICAL ANALYSIS?

The ability to identify heterozygosity in any genetically transmitted disease is of paramount importance for genetic counseling. Laron et al.[54] evaluated the correlation between biochemical parameters in 13 patients and their 16 family members to determine if heterozygosity could be determined. The study was based on the presumed obligate heterozygosity of the parents in lieu of molecular analysis. The data suggested that 70% of presumed heterozygotes can be identified by analysis of GHBP activity in sera.

Subsequently, Fielder et al.[55] evaluated biochemical parameters of Ecuadorian probands who were homozygous for the same point mutation of the GH receptor gene, 17 parents with proven heterozygosity, and 21 healthy Ecuadorians. They found no significant differences of serum IGF-I and IGFBP-3 levels, and GHBP binding capacity between heterozygotes and controls. Insulin-like growth factor-II was modestly lower in the mothers as compared to controls. Taken together, these data indicate that only molecular analysis will adequately identify the heterozygote state.

V. MOLECULAR ANALYSIS OF THE GROWTH HORMONE RECEPTOR GENE

The GH receptor gene has been localized to chromosome 5p13.1-p12 and spans >87 kb.[56] The gene contains 620 amino acids within three separate domains: the extracellular domain which is structurally identical to the GHBP (246 amino acids), the cytoplasmic domain (350 amino acids), and a transmembrane domain which separates the two. Human and rabbit GHBP have an amino terminal sequence identical to the membrane-associated GH receptor and are presumed to derive from proteolytic cleavage of the extracellular domain. Further characterization of the GH receptor gene demonstrated that exon 2 encodes the signal peptide, exon 3 through 7 the extracellular domain, exon 8 the transmembrane domain, and exons 9 and 10 the cytoplasmic domain.[18] The mechanism for signal transduction of the GH receptor is unknown.

Godowski et al.[18] initially described, in two of nine patients, deletions of the GH receptor gene that were consistent with exons 4, 5, and 6 on DNA restriction patterns. Further analysis demonstrated the failure of the DNA from these subjects to hybridize to probes for exons 3, 5, and 6 and the production of an abnormal size band when hybridized with probes for exon 4. These deletions affect large portions of the extracellular domain, and include but are not limited to the region specific for GH binding.

Amselem and colleagues[19] were the first to identify a point mutation within the extracellular domain of the GH receptor gene. In four Tunisian siblings of consanguineous parents, a T–C substitution in codon 96 of exon 5 was identified which generated a Ser in place of a Phe. Bass and Wells developed a mutant protein in *Escherichia coli* by oligonucleotide-directed mutagenesis consistent with this mutation. The binding affinity for GH of the mutant, however, was the same as the wild-type GHBP.[57] Duquesnoy and colleagues[58] argued that the bacterial expression of such a protein may lack post-translational modifications characteristic of the hGH receptor. They expressed the hGH receptor complementary DNA and the mutant protein in mammalian COS-7 monkey kidney cells and found that [^{125}I]-hGH binding was limited to cellular lysosomal fractions and was undetectable only in mutant plasma membranes. Edery and co-workers also used COS-7 cells for transient expression of the wild-type GH receptor and GHBP and their mutants. They found that all mutant forms were incapable of binding hGH.[59]

Two nonsense mutations were also described in patients with low to absent GHBP activity.[60] A substitution of C-T at nucleotide 181 resulting in the loss of most of the GH receptor with termination of the protein at codon 43 of exon 4 was found in two separate consanguineous families. Also, a C-A substitution at nucleotide 168 resulted in termination at codon 38 of exon 4.

The discovery of a high concentration of affected individuals in southern Ecuador presented Berg et al.[25] with a unique opportunity to identify the GH receptor gene mutation in a group that was ethnically homogeneous. Analysis of the GH receptor gene from an obligate heterozygote by polymerase chain reaction amplification and denaturing gradient gels electrophoresis identified changes in exon 6 consistent with a mutation. A single nucleotide substitution of an adenine to guanine (GAA-GAG) in the third position (nucleotide 594) of codon 180 was identified by sequencing. This substitution did not change the normally encoded amino acid, glutamic acid. Further analysis, however, demonstrated a deletion of 24 nucleotides encoding 8 amino acids (residues 181 to 188) from the extracellular domain at exon 6. The new nucleotide sequence, which shares homology with the 5′ splice consensus sequence, creates a mutated splice site incapable of coding for residues 181 to 188. They hypothesized that the deletion may affect the folding of the protein, making it highly sensitive to rapid degradation. This point mutation was identified in 55 of 56 Ecuadorian patients, making this the largest genetically homogeneous group reported thus far.[25,32] With the use of allele-specific oligonucleotide hybridization the codon 180 mutation was detected in all Ecuadorian probands, their parents, and two thirds of their siblings, while the mutant allele was not detected in 61 controls.[25] The one Ecuadorian patient that did not share homozygosity for this substitution was homozygous for a nonsense mutation previously detected in two Mediterranean families.[19,25]

Eighteen separate point mutations of the GH receptor gene were reported and include three nonsense mutations, nine missense mutations, four splice junction mutations, and two frameshift mutations (Figure 1).[32,60] Two of the 18 identified mutations were present in one patient as single alleles inherited from the mother and involved the cytoplasmic domain. The patient, although known to have decreased GHBP activity, responded favorably to GH therapy.[60] Further investigation of this case is warranted.

VI. THERAPEUTIC POTENTIAL OF RECOMBINANT INSULIN-LIKE GROWTH FACTOR-I

Pharmacological studies of IGF-I in healthy subjects were initially designed to evaluate safety and metabolic effects after intravenous bolus or constant infusion.[14,15] Guler et al. compared the metabolic effects of intravenous IGF-I to insulin in healthy adults and noted a reduction of blood glucose levels (>50% of baseline) 30 min after a bolus injection (100 μg/kg), and normalization within 2 h.[14] This effect was attributed to the increase in free IGF-I noted by 15 min postinjection (19% pretreatment to 80%). The hypoglycemic potency of IGF-I was determined to be only 6% that of insulin on an equimolar basis. In contrast, IGF-I administered over 6 d by constant intravenous infusion (20 μg/h), did not

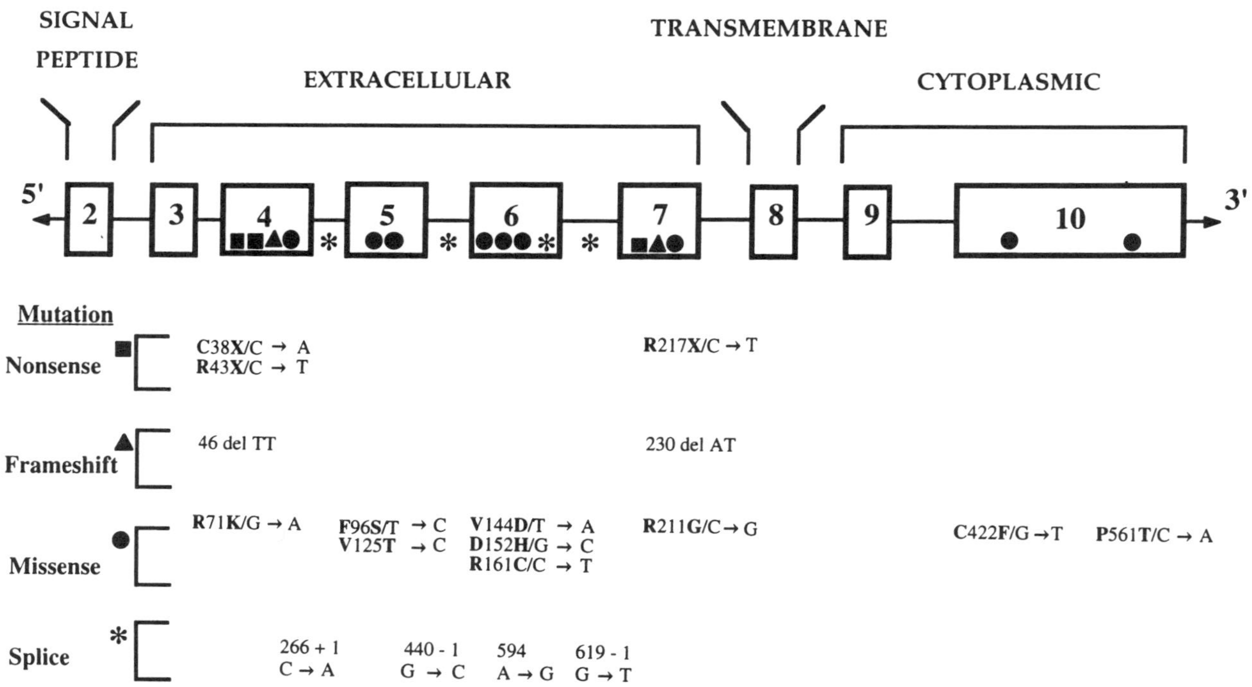

FIGURE 1. Diagrammatic representation of the GH receptor gene and the 18 reported mutations found to date in patients with Laron syndrome. Boxes represent the nine coding exons (2 to 10). The mutations are identified by a ■ (nonsense), ▲ (frameshift), ● (missense), or ✱ (splice), and are found within the respective exon. Nonsense and missense mutations are described by bold letters which flank the codon number and correspond to the expected and the mutant amino acid. Frameshift mutations are represented by the codon number and the dinucleotide deletion. Splice mutations are described by the nucleotide number and the numerical representation of the shift created in the splice site relative to the respective exon. All nucleotide substitutions are depicted by an arrow.

produce hypoglycemia even though serum IGF-I levels increased more than fourfold (150 to 700 μg/l).[61] Protein synthesis and creatinine clearance increased, while serum GH levels were suppressed.

The pharmacodynamics of IGF-I and -II administered by subcutaneous bolus injection to healthy adults were subsequently examined with the use of radiolabeled IGFs.[62] After IGF-I injection, serum IGF-II levels decreased concomitant with the increase of serum IGF-I levels. The half-life of free IGF-I and -II was 10 to 12 min and the production rate was 10 to 13 mg/d. Bound IGFs initially migrated with the 50 kDa binding protein complex, maintaining a half-life of 20 to 30 min and subsequently with the 200 kDa complex with which IGF half-life reached 12 to 15 h.

Laron et al.[15] were the first to administer IGF-I to GHRD subjects; nine patients (11 to 33 years) received intravenous bolus injections (75 μg/kg) after a 10-h fast. Prolonged symptomatic hypoglycemia (45% of baseline) developed by 30 min and required that a meal be given after 2 h for resolution of symptoms. It was hypothesized that the acuteness and severity of hypoglycemia seen in this group was due to a sustained increase in free IGF-I resulting from inadequate IGFBP-3 in serum. Increased levels of GH in serum were presumed to be in response to acute hypoglycemia. Sustained hypoglycemia concomitant with elevated GH levels suggested that the inability of GH to produce a counterregulatory effect was due to the GH receptor defect.

Klinger et al.[63] compared the distribution and clearance characteristics of IGF-I given by intravenous bolus (75 μg/kg) to ten patients with LS and six healthy volunteers. The half-life of IGF-I was significantly shorter in LS patients (2.6 ± 0.7 h vs. 4.4 ± 0.5h, $p < 0.01$), elimination more rapid, and distribution half-life was similar for both groups. Rapid elimination of IGF-I in LS patients was attributed to the lack of IGFBP-3.

Walker and colleagues[64] observed significant anabolic effects and asymptomatic hypoglycemia during constant intravenous infusion of IGF-I to a 9-year-old boy over 11 d. Serum IGFBP-2 levels increased significantly (164 ± 33 to 580 μg/l) during treatment and IGFBP-3 was not reported.[65] Subsequent clinical trials in GHRD patients employed subcutaneous delivery of IGF-I to reduce the risk of hypoglycemia and facilitate long-term administration of IGF-I.

The pharmacokinetics and metabolic effects of recombinant IGF-I were evaluated in six Ecuadorian adults with GHRD after the subcutaneous injection of IGF-I (40 μg/kg every 12 h) for 7 d.[46] With regular meals and snacks, blood glucose levels reached a nadir 3 to 5 h postinjection, but without hypoglycemia. Fasting insulin levels were unchanged in contrast to 2 h postprandial values, which were suppressed. A twofold increase in urinary excretion of calcium was observed concomitant with unchanged serum calcium levels. Anabolic effects were not observed. Comparison of 24-h GH profiles obtained before and during IGF-I administration demonstrated significant IGF-I suppression of all parameters measured. Serum IGF-I levels achieved a peak 2 to 6 h postinjection (mean 253 ± 11

TABLE 3
Effect of Long-Term IGF-I Therapy on Growth Velocity in GHIS

No. of patients/Dx	Age (years)	Duration (months)	Dosage (μg/kg)	Growth velocity (cm/year) Pre-Rx	Growth velocity (cm/year) Rx	Ref.
5/LS	3–14	3–10	150/d	2.8–5.8	8.8–13.6	26
1/LS	9	9	120 BID	6.5	11.4	27
5/LS	3–22	6	40 BID	4.1	8.2	29
7/LS	3–22	9	40–120 BID	4.1	7.1	29
17/LS	3–22	6	120 BID	4.7	10.2	29
2/GHD-1A	14, 16	6	120 BID	1.4	8.4	29
1/LS	17	12	120 BID	4.1[a]	7.8	30
1/LS	18	12	120 BID	3.4	8.0	30
1/LS	10	9	40–120 BID	3.6	10.2	31
5/LS 3/GHD-1A	2–11	3–18	80–120 BID	4.2	10.6	28

Abbreviations: GHIS, growth hormone insensitivity syndrome; LS, Laron syndrome; BID, twice daily; and GHD-1A, GH deficiency type 1A.

[a]Patient was receiving GnRH analogue during the time period used for the calculation of growth velocity.

μg/l) and trough levels (mean 137 ± 8 μg/l) remained above the pretreatment mean. These values were within the range seen for healthy Ecuadorian controls. Serum IGF-II levels decreased concomitant with the increase in serum IGF-I. The IGFBP-3 levels were unchanged, while serum IGFBP-2 increased significantly.

The efficacy of long-term IGF-I therapy for children with LS who have extremely low levels of IGFBP-3 had been speculative. There were two major concerns; expected elevated levels of free IGF-I could increase the risk for hypoglycemia, and without normalization of IGFBP-3 levels in serum, reduced IGF-I half-life would diminish the beneficial effects of IGF-I. Studies designed to evaluate the relationship between IGFBP-3 and IGF-I production after intravenous GH administration to GH-deficient patients indicate that serum IGFBP-3 levels increase only after IGF-I increases.[66] This suggested that IGFBP-3 production was directly regulated by IGF-I and that long-term IGF-I therapy would stimulate IGFBP-3 synthesis in LS patients and normalize IGF-I pharmacokinetics.

The first observations of long-term IGF-I therapy were reported in 1992. Laron et al.[26] treated five children and noted an increase in growth velocities from 2.8 to 5.8 to 8.8 to 13.6 cm/year (Table 3). Similarly, one patient's growth velocity increased after 9 months of IGF-I therapy, and patients with GH deficiency type IA also demonstrated catch-up growth during IGF-I treatment (Table 3).[27–29]

The largest IGF-I therapeutic trial conducted thus far includes 30 LS children from Europe and Australia (3 to 23 years).[29] Several responded poorly to initial

TABLE 4
Reported Side Effects with IGF-I Therapy

Tachycardia	3
Hypoglycemia (<3 mmol/l)	17
Hypokalemia	2
Hypotension	1
Seizures (with hypoglycemia)	2
Headache	12
Ureterolithiasis	1
Papilloedema	3
Right facial nerve paresis	1
Increased transaminases	2
Thrombocytosis	1

IGF-I therapy (40 μg/kg twice daily) with increases in growth velocity <2 cm/year above baseline, prompting an increase in dosage to 60 or 120 μg/kg twice daily (Table 3). Overall, increased growth velocity was achieved by most patients and a dose-dependent response was observed. Two failures were reported in the eldest patients (20 and 23 years) and were probably related to completion of puberty. Surprisingly, IGFBP-3 levels were unchanged during therapy. Hypoglycemia occurred in 17 patients, 2 of whom developed seizures, and 1 had papillodema on examination (Table 4). Another patient experienced hypotension, hypokalemia, and fatigue shortly after each injection, causing him to withdraw from the study.

Several IGF-I therapeutic trials are currently underway in Ecuador for GHRD. Two peripubertal patients received a long-acting GnRH analogue, as well as IGF-I, and demonstrated sustained growth without an increase in serum IGFBP-3 (Table 3).[30] A randomized, double-blind, placebo-controlled study involving IGF-I therapy, involving 17 GHRD children, is being conducted in Ecuador.

Clinical trials demonstrating the growth-promoting potential of IGF-I in LS have been encouraging. Little is known relative to the optimal dose and frequency of administration, but it is clear that IGF-I does not stimulate IGFBP-3 synthesis in the presence of a GH receptor defect.[29,30,46,67] Twice-daily dosing of IGF-I has been employed in an attempt to sustain the steady-state kinetics of IGF-I in a system devoid of IGFBP-3.[46] One study, however, reported good growth velocity in response to daily injections.[26] The failure of IGF-I to enhance IGFBP-3 production while stimulating a significant increase in growth velocity suggests that IGF-I half-life and IGFBP-3 serum levels do not reflect the end-organ interaction between ligand and receptor. In contrast, the bioavailability of IGF-I may be enhanced by low serum levels of IGFBP-3. The concomitant rise of IGFBP-2, which is more readily accessible to the interstitial space than IGFBP-3, may also contribute to the bioavailability of IGF-I.[46,62,65]

Although the growth-promoting effects of IGF-I were confirmed, many adverse events have been reported since the institution of long-term therapy (Table 4).

Further trials are necessary to determine the most effective dosage of IGF-I that does not produce toxicity and whether the growth response will be sustained.

VII. SUMMARY

The advances that have been made in the past 25 years regarding the clinical, biochemical, and genetic characterization, and therapy of LS have also increased our knowledge of the physiology of growth. Many questions, however, remain unanswered. How are GHBP and IGFBP-3 regulated and what is their function? What is the best biochemical indicator of an adequate dosage of IGF-I? Are factors which are seen consistently with enhanced growth during IGF-I therapy, such as increased urinary calcium excretion, suppression of GH release, or increased IGFBP-2, useful indicators? Do the moderate increases in binding proteins other than IGFBP-3 improve the bioavailability of exogenous IGF-I, and if so, how can this be maximized? There is increasing evidence of a direct stimulatory effect of IGF-I on IGFBP-2.[68,69] Whether IGFBP-2 is directly responsible for enhanced IGF-I bioavailability in LS remains to be determined.

Current IGF-I therapeutic trials for children with LS may help to answer many of these questions. The results thus far have been promising, as recombinant IGF-I appears to ameliorate the most stigmatizing clinical feature characteristic of this condition, severe short stature. Whether IGF-I therapy can be of therapeutic benefit to adults with GHRD will depend on observations made from studies designed to determine whether the possible increased risk for coronary heart disease and early osteoporosis might be decreased by IGF-I therapy.[70]

ACKNOWLEDGMENT

The author wishes to acknowledge the following colleagues for their contributions: Jaime Guevara-Aguirre, M.D.; Ron G. Rosenfeld, M.D.; Uta Francke, M.D.; Mary Anne Berg, M.D.; Paul Fielder, Ph.D.; Frank B. Diamond, M.D.; Pinchas Cohen, M.D.; and Sharron Gargosky, Ph.D. The author is especially indebted to Arlan L. Rosenbloom, M.D., and Kathleen T. Shiverick, Ph.D., for their mentorship and critical review of this manuscript. This work was supported by NRSA grant HD07595–02.

REFERENCES

1. **Laron, Z., Pertzelan, A., and Mannheimer, S.,** Genetic pituitary dwarfism with high serum concentration of growth hormone: a new inborn error of metabolism?, *Isr. J. Med. Sci.*, 2, 152, 1966.
2. **Laron, Z., Blum, W., Chatelain, P., Ranke, M., Rosenfeld, R., Savage, M., and Underwood, L.,** Classification of growth hormone insensitivity syndrome, *J. Pediatr.*, 122, 241, 1993.

3. **Guevara-Aguirre, J., Rosenbloom, A. L., Vaccarello, M. A., Fielder, P. J., de la Vega, A., Diamond, F. B., Jr., and Rosenfeld, R. G.,** Growth hormone receptor deficiency (Laron syndrome): clinical and genetic characteristics, *Acta Paediatr. Scand.*, 377, 96, 1991.
4. **Laron, Z., Pertzelan, A., and Karp, M.,** Pituitary dwarfism with high serum levels of growth hormone, *Isr. J. Med. Sci.*, 4, 883, 1968.
5. **Merimee, T. J., Hall, J., Rabinovitz, D., McKusick, V. A., and Rimoin, D. L.,** An unusual variety of endocrine dwarfism: subresponsiveness to growth hormone in a sexually mature dwarf, *Lancet*, 2, 191, 1968.
6. **Daughaday, W. H., Laron, Z., Pertzelan, A., and Heins, J. N.,** Defective sulfation factor generation: a possible etiological link in dwarfism, *Trans. Assoc. Am. Phys.*, 82, 129, 1969.
7. **Bala, R. M. and Beck, J. C.,** Fractionation studies on plasma of normals and patients with Laron dwarfism and hypopituitary gigantism, *Can. J. Physiol. Pharmacol.*, 91, 845, 1973.
8. **Eshet, R., Laron, Z., Brown, M., and Arnon, R.,** Immunoreactive properties of the plasma hGH from patients with the syndrome of familial dwarfism and high IR-hGH, *J. Clin. Endocrinol. Metab.*, 37, 819, 1973.
9. **Jacobs, L. S., Sneid, D. S., Garland, J. T., Laron, Z., and Daughaday, W. H.,** Receptor-active growth hormone in Laron dwarfism, *J. Clin. Endocrinol. Metab.*, 42, 403, 1976.
10. **Eshet, R., Laron, Z., Pertzelan, A., Arnon, R., and Dintzman, M.,** Defect of human growth hormone receptors in the liver of two patients with Laron-type dwarfism, *Isr. J. Med. Sci.*, 20, 8, 1984.
11. **Niwa, M., Sato, Y., Uchiyamo, F., Ono, H., Yamashita, M., and Kitaguchi, T.,** Chemical synthesis cloning and expression of genes for human somatomedin-C (insulin-like growth factor I) and val-somatomedin-C, *Ann. N.Y. Acad. Sci.*, 469, 31, 1986.
12. **Geffner, M. E., Golde, D. W., Lippe, B. M., Kaplan, S. A., Bersche, N., and Li, C. H.,** Tissues of Laron dwarfs are sensitive to insulin-like growth factor I but not to growth hormone, *J. Clin. Endocrinol. Metab.*, 64, 1042, 1987.
13. **Guler, H. P., Zaph, J., Schweiwiller, E., and Froesch, E. R.,** Recombinant human insulin-like growth factor-I stimulates and has distinct effects on organ size in hypophysectomized rats, *Proc. Natl. Acad. Sci. U.S.A.*, 85, 4889, 1988.
14. **Guler, H. P., Zapf, J., and Froesch, E. R.,** Short-term metabolic effects and half-lives of intravenously administered insulin-like growth factor I in healthy adults, *N. Engl. J. Med.*, 317, 137, 1987.
15. **Laron, Z., Erster, B., Klinger, B., and Anin, S.,** Effects of acute administration of insulin-like growth factor-I in patients with Laron-type dwarfism, *Lancet*, 2, 1170, 1988.
16. **Daughaday, W. H. and Trivedi, B.,** Absence of serum growth hormone binding protein in patients with growth hormone receptor deficiency (Laron dwarfism), *Proc. Natl. Acad. Sci. U.S.A.*, 84, 4636, 1987.
17. **Leung, D. W., Spencer, S. A., Cachianes, G., Hammonds, R. G., Collins, C., Henzel, W. J., Barnard, R., Waters, M. J., and Wood, W. I.,** Growth hormone receptor and serum binding protein: purification, cloning, and expression, *Nature*, 330, 537, 1987.
18. **Godowski, P. J., Leung, D. W., Meacham, L. R., Galgani, J. P., Hellmiss, R., Keret, R., Rotwein, P. S., Parks, J. S., Laron, Z., and Wood, W. I.,** Characterization of the human growth hormone receptor gene and demonstration of a partial gene deletion in two patients with Laron-type dwarfism, *Proc. Natl. Acad. Sci. U.S.A.*, 86, 8083, 1989.
19. **Amselem, S., Duquesnoy, B. S., Attree, O., Novelli, G., Bousnina, S., Postel-Vinay, M. C., and Goossens, M.,** Laron dwarfism and mutations of the growth hormone-receptor gene, *N. Engl. J. Med.*, 321, 989, 1989.
20. **Rosenbloom, A. L., Guevara-Aguirre, J., Rosenfeld, R. G., and Fielder, P. J.,** The little woman of Loja: growth hormone receptor deficiency in an inbred population of Southern Ecuador, *N. Engl. J. Med.*, 323, 1367, 1990.
21. **Buchanan, C. R., Maheshwari, H. G., Normal, M. R., Morrell, D. J., and Preece, M. A.,** Laron-type dwarfism with apparently normal high affinity serum growth hormone-binding protein, *Clin. Endocrinol.*, 35, 179, 1991.

22. **Guevara-Aguirre, J., Rosenbloom, A. L., Fielder, P. J., Diamond, F. B., Jr., and Rosenfeld, R. G.,** Growth hormone receptor deficiency in Ecuador: clinical and biochemical phenotype in two populations, *J. Clin. Endocrinol. Metab.*, 76, 417, 1993.
23. **Cunningham, B. C., Usch, M., and Kossiakoff, A.,** Dimerization of the extracellular domain of the human growth hormone receptor by a single hormone molecule, *Science,* 254, 821, 1991.
24. **De Vos, A. M., Ultsch, M., and Kossiakoff, A. A.,** Human growth hormone and extracellular domain of its receptor: crystal structure of the complex, *Science,* 255, 306, 1992.
25. **Berg, M. A., Guevara-Aguirre, J., Rosenbloom, A. L., Rosenfeld, R. G., and Francke, U.,** Mutation creating a new splice site in the growth hormone receptor genes of 37 Ecuadorean patients with Laron syndrome, *Hum. Mutat.*, 1, 24, 1992.
26. **Laron, Z., Anin, S., Klipper-Auerbach, Y., and Klinger, B.,** Effects of insulin-like growth factor-I on linear growth, head circumference, and body fat in patients with Laron-type dwarfism, *Lancet,* 339, 1258, 1992.
27. **Walker, J., Van Wyk, J. J., and Underwood, L. E.,** Stimulation of statural growth by recombinant insulin-like growth factor-I in a child with growth hormone insensitivity syndrome (Laron type), *J. Pediatr.*, 121, 641, 1992.
28. **Backeljauw, P. F. and Underwood, L. E.,** Effects of prolonged IGF-I treatment in children with growth hormone insensitivity syndrome (GHIS), *Pediatr. Res.*, 33, S56, 1993.
29. **Savage, M. O., Wilton, P., Ranke, M. B., Chatelain, P. G., Blum, W. F., Cotterill, A. M., Preece, M. A., and Rosenfeld, R. G.,** Therapeutic response to recombinant IGF-I in thirty two patients with growth hormone insensitivity, *Pediatr. Res.*, 33, S5, 1993.
30. **Martinez, V., Vasconez, O., Martinez, A., Moreno, Z., Davila, N., Rosenbloom, A. L., Diamond, F. B., Bachrach, L., Rosenfeld, R. G., and Guevara-Aguirre, J.,** Body changes in adolescent growth hormone receptor deficient patients receiving rhIGF-I and a LRH analogue: preliminary results, *Acta Paediatr. Suppl.*, 399, 133, 1994.
31. **Martha, P. M., Jr., Johnston, C., Cohen, P., Rosenfeld, R., and Reiter, E. O.,** Short term metabolic changes and long term response to recombinant IGF-I (rhIGF-I) therapy in a North American child with GH insensitivity, *Pediatr. Res.*, 33, S49, 1993.
32. **Berg, M. A., Perez-Jurado, L., Guevara-Aguirre, J., Rosenbloom, A. L., Laron, Z., Milner, R. D. G., and Francke, U.,** Receptor mutations in growth hormone insensitivity syndrome (GHIS): a global survey and identification of the Ecuadorean E180 splice mutation in an Oriental Jewish patient, *Acta Paediatr. Suppl.*, 399, 112, 1994.
33. **Rosenfeld, R. G., Rosenbloom, A. L., and Guevara-Aguirre, J.,** Growth hormone insensitivity due to growth hormone receptor deficiency, *Endocr. Rev.*, 15, 369, 1994.
34. **Laron, Z.,** Laron-type dwarfism (hereditary somatomedin deficiency), in *Advances in Internal Medicine and Pediatrics,* Frick, P., von Harnack, G. A., Kochsieck, K., Martini, G. A., and Prader, A., Eds., Springer-Verlag, Berlin, 1984, 117.
35. **Rivarola, M. A., Phillips, J. A., III, Migeon, C. J., Heinrich, J. J., and Hjelle, B. J.,** Phenotypic heterogeneity in familial isolated growth hormone deficiency type I-A, *J. Clin. Endocrinol.*, 59, 34, 1984.
36. **Schaefer, G. B., Rosenbloom, A. L., Guevara-Aguirre, J., Campbell, E. A., Ulrich, F., Patil, K., and Frias, J. L.,** Facial morphometry of Ecuadorian patients with growth hormone receptor deficiency Laron syndrome, *J. Med. Gen.*, 31, 365, 1994.
37. **Laron, Z.,** Laron syndrome: from description to therapy, *Endocrinologist,* 3, 21, 1993.
38. **Laron, Z., Lilos, P., and Klinger, B.,** Growth curves for Laron syndrome, *Arch. Dis. Child.*, 68, 768, 1993.
39. **Savage, M. O., Chatelain, P. G., Preece, M. A., Ranke, M. B., Sietnieks, A., and Wilton, P.,** Clinical spectrum of the syndrome of growth hormone insensitivity, *Acta Paediatr. Scand. Suppl.*, 377, 87, 1991.
40. **Laron, Z., Sarel, R., and Pertzelan, A.,** Puberty in Laron type dwarfism, *Eur. J. Pediatr.*, 134, 79, 1980.

41. **Menashe, Y., Sack, J., and Mashiach, S.,** Spontaneous pregnancies in two women with Laron-type dwarfism: are growth hormone and circulating insulin-like growth factor mandatory for induction of ovulation?, *Hum. Reprod.*, 6, 670, 1991.
42. **Guevara-Aguirre, J. and Rosenbloom, A. L.,** Psychosocial adaptation of Ecuadorian patients with growth hormone receptor deficiency/Laron syndrome, in *Lessons from Laron Syndrome (LS) 1966–1992,* No. 24, Laron, Z. and Parks, J. S., Eds., S. Karger, Basel, 1993, 61.
43. **Galatzer, A., Aran, O., Nagelberg, N., Rubitzek, J., and Laron, Z.,** Cognitive and psychosocial functioning of young adults with Laron syndrome, in *Lessons from Laron Syndrome (LS) 1966–1992,* No. 24, Laron, Z. and Parks, J. S., Eds., S. Karger, Basel, 1993, 53.
44. **Berelowitz, M., Szabo, M., Frohman, L. A., Firestone, S., and Chu, L.,** Somatomedin-C mediates growth hormone negative feedback by effects on both the hypothalamus and the pituitary, *Science,* 212, 1279, 1981.
45. **Keret, R., Pertzelan, A., Zeharia, A., Zadik, Z., and Laron, Z.,** Growth hormone (hGH) secretion and turnover in three patients with Laron-type dwarfism, *Isr. J. Med. Sci.*, 24, 75, 1988.
46. **Vaccarello, M. A., Diamond, F. B., Jr., Guevara-Aguirre, J., Rosenbloom, A. L., Fielder, P. J., Gargosky, S., Cohen, P., Wilson, K., and Rosenfeld, R. G.,** Hormonal and metabolic effects and pharmacokinetics of recombinant insulin-like growth factor-I in growth hormone receptor deficiency (GHRD)/Laron syndrome, *J. Clin. Endocrinol. Metab.*, 77, 273, 1993.
47. **Cohen, P., Fielder, P. J., Hasegawa, Y., Frish, H., Giudice, L. C., and Rosenfeld, R. G.,** Clinical aspects of insulin-like growth factor binding proteins, *Acta Endocrinol.*, 124 (Suppl.), 74, 1991.
48. **Hardouin, S., Gourmelen, M., Noguiez, P., Seurin, D., Roghani, M., Le Bouc, Y., Povoa, G., Merimee, T. J., Hossenlopp, P., and Binoux, M.,** Molecular forms of serum insulin-like growth factor (IGF)-binding proteins in man: relationships with growth hormone and IGFs and physiological significance, *J. Clin. Endocrinol. Metab.*, 69, 1291, 1989.
49. **Glasscock, G. F., Hein, A. N., Miller, J. A., Hintz, R. L., and Rosenfeld, R. G.,** Effects of continuous infusion of insulin-like growth factor I and II, alone and in combination with thyroxine or growth hormone, on the neonatal hypophysectomized rat, *Endocrinology,* 130, 203, 1992.
50. **Wilson, K. F., Fielder, P. J., Guevara-Aguirre, J., Rosenbloom, A. L., and Rosenfeld, R. G.,** Long Term Effects of IGF-I Therapy on the IGF-IGFBP Axis in Growth Hormone Receptor Deficiency (GHRD), Abstr. 1599A, presented at 75th Annu. Meet. Endocrine Society, Las Vegas, Nevada, 1993.
51. **Baumann, G., Stolar, M. W., Amburn, K., Barsano, C. P., and DeVries, B. C.,** A specific growth hormone-binding protein in human plasma: initial characterization, *J. Clin. Endocrinol. Metab.*, 62, 134, 1986.
52. **Carlsson, L. M. S., Rowland, A. M., Clark, R. G., Gesundheit, N., and Wong, W. L. T.,** Ligand-mediated immunofunctional assay for quantitation of growth hormone-binding protein in human blood, *J. Clin. Endocrinol. Metab.*, 73, 1216, 1991.
53. **Savage, M. O., Blum, W. F., Ranke, M. B., Postel-Vinay, M. C., Cotterill, A. M., Hall, K., Chatelain, P. G., Preece, M. A., Rosenfeld, R. G.,** Clinical features and endocrine status in patients with growth hormone insensitivity (Laron syndrome), *J. Clin. Endocrinol. Metab.*, 77, 1465–1471, 1993.
54. **Laron, Z., Klinger, B., Erster, B., and Silbergeld, A.,** Serum GH binding protein activities identifies the heterozygous carriers for Laron type dwarfism, *Acta Endocrinol. (Copenhagen),* 121, 603, 1989.
55. **Fielder, P. J., Guevara-Aguirre, J., Rosenbloom, A. L., Carlsson, L., Hintz, R. L., and Rosenfeld, R. G.,** Expression of serum insulin-like growth factors, insulin-like growth factor-binding proteins, and the growth hormone-binding protein in heterozygote relatives of Ecuadorian growth hormone receptor deficient patients, *J. Clin. Endocrinol. Metab.*, 74, 743, 1992.
56. **Barton, D. E., Foellmer, B. E., Wood, W. I., and Francke, U.,** Chromosome mapping of the growth hormone receptor gene in man and mouse, *Cytogen. Cell Gen.*, 50, 137, 1989.

57. **Bass, S. and Wells, J.,** Growth hormone-receptor gene in Laron dwarfism, *N. Engl. J. Med.,* 322, 854, 1990.
58. **Duquesnoy, P., Sobrier, M. L., Amselem, S., and Goossens, M.,** Defective membrane expression of human growth hormone (GH) receptor causes Laron-type GH insensitivity syndrome, *Proc. Natl. Acad. Sci. U.S.A.,* 88, 10272, 1991.
59. **Edery, M., Rozakis-Adcock, M., Goujon, L., Finidori, J., Levy-Meyrueis, C., Paly, J., Dijane, J., Postel-Vinay, M.-C., and Kelly, P. A.,** Lack of hormone binding in COS-7 cells expressing a mutated growth hormone receptor found in Laron dwarfism, *J. Clin. Invest.,* 91, 838–844, 1993.
60. **Amselem, S., Duquesnoy, P., Duriez, B., Dastot, F., Sobrier, M.-L., and Valleix, S.,** Spectrum of growth hormone receptor mutations and associate haplotypes in Laron syndrome, *Hum. Mol. Gen.,* 4, 355, 1993.
61. **Guler, H. P., Schmid, C., Zapf, J., and Froesch, R.,** Effects of recombinant insulin-like growth factor-I on insulin secretion and renal function in normal human subjects, *Proc. Natl. Acad. Sci. U.S.A.,* 86, 2868, 1989.
62. **Guler, H. P., Zapf, J., Schmid, C., and Froesch, E. R.,** Insulin-like growth factors I and II in healthy man. Estimations of half-lives and production rates, *Acta Endocrinol. (Copenhagen),* 121, 753, 1989.
63. **Klinger, B., Garty, M., and Laron, Z.,** Elimination characteristics of intravenously administered rIGF-I in Laron-type dwarfs, *Dev. Pharmacol. Ther.,* 15, 196, 1990.
64. **Walker, J. L., Ginalska-Malinowska, M., Romer, T. E., Pucilowska, J. B., and Underwood, L. E.,** Effects of the infusion of insulin-like growth factor I in a child with growth hormone insensitivity syndrome (Laron dwarfism), *N. Engl. J. Med.,* 324, 1483, 1991.
65. **Walker, J. L., Baxter, R. C., Young, S. C. J., Pucilowska, J. B., and Underwood, L. E.,** Effects of recombinant insulin-like growth factor I (IGF-I) on IGF binding proteins (IGF-BPS) and the acid labile subunit (ALS) in growth hormone insensitivity syndrome (GHIS), Abstr. 554, presented at 74th Annu. Meet. Endocrine Society, 1992.
66. **Jorgensen, J. O. L., Blum, W. F., Moller, N., Ranke, M. B., and Christiansen, J. S.,** Short-term changes in serum insulin-like growth factors (IGFs) and IGF-binding protein-3 after different modes of intravenous growth hormone (GH) exposure to GH deficient patients, *J. Clin. Endocrinol. Metab.,* 72, 582, 1991.
67. **Laron, Z., Klinger, B., Blum, W. F., Silbergeld, A., and Ranke, M. B.,** Effect of IGF-I on IGFBP-3 in Laron dwarfism, *Clin. Endocrinol.,* 36, 301, 1992.
68. **Boulware, S. D., Rennert, N. J., Busby, W. H., Tamborlane, W. V., and Sherwin, R. S.,** Differential responsiveness of IGF binding proteins (IGFBP)-1 and 2 to IGF-1 and insulin (INS) infusions in humans, Abstr. No. 1600C, presented at 75th Annu. Meet. Endocrine Society, Abstr. 1993.
69. **Zenobi, P. D., Schwander, J., Zapf, J., and Froesch, E. R.,** Recombinant human insulin-like growth factor (RHIGF)-I treatment in type 2 diabetes mellitus alters the profile of IGF-binding proteins (IGFBPs), Abstr. No. 558, presented at 74th Annu. Meet. Endocrine Society, 1992.
70. **Corpas, E., Mitchell, H., and Blackman, M. R.,** Human growth hormone and human aging, *Endocr. Rev.,* 14, 20, 1993.

Chapter 8

GROWTH HORMONE, INSULIN-LIKE GROWTH FACTOR-I, AND HUMAN AGING

Robert Marcus and Andrew Hoffman

TABLE OF CONTENTS

I. INTRODUCTION

Normal human aging is characterized by well-documented changes in body composition and somatotropic function. The parallel and apparently contemporaneous nature of these events has led to the attractive hypothesis that the changes in composition result to some growth hormone-insulin-like growth hormone (GH-IGF-I) axis.[1] If this hypothesis is true, restitution of youthful levels of GH and IGF-I might then reverse the presumably undesirable somatic features of advancing age. Until recently, limited availability and extraordinary cost of human pituitary GH rendered such an experiment impractical. The development of recombinant human GH (rhGH) now makes it reasonable to consider the therapeutic use of rhGH in elderly men and women if potential benefits can be established and if side effects and long-term health risks can be avoided or minimized. In this chapter we review age-related changes in the somatotropic axis and in body composition in humans and describe the results of studies using rhGH in older people. The basic physiology and molecular endocrinology of the somatotropic axis are discussed in other chapters in this volume and are considered here only by way of introduction.

II. ORGANIZATION OF THE SOMATOTROPIC AXIS

Synthesis and secretion of GH by the somatotropic cells of the anterior pituitary is tightly regulated by multiple hormones and neurotransmitters. The proximate stimulus for GH secretion, GH releasing hormone (GHRH), is produced in the hypothalamus and carried via the portal circulation to the anterior pituitary, where it interacts with specific cell-surface receptors on somatotropes to promote the pulsatile secretion of GH. By contrast, local concentrations of somatostatin inhibit GH release. Growth hormone interacts with receptors in liver and peripheral tissues to stimulate the production and release of IGF-I. Circulating IGF-I derives primarily from the liver, but GH stimulates local production of IGF-I by at least some peripheral tissues, leading in an autocrine or paracrine fashion to the typical physiological responses associated with GH action. Insulin-like growth factor directly interacts with type I IGF receptors on somatotropes to inhibit GH secretion, thereby completing a classical negative feedback loop for this system. In addition, IGF-I can inhibit GHRH release and stimulate somatostatin secretion in the brain and GH may inhibit its own production by increasing somatostatinergic tone.[2]

III. AGE-RELATED CHANGES IN THE SOMATOTROPIC AXIS AND BODY COMPOSITION

A. THE SOMATOTROPIC AXIS

Growth hormone secretion declines during normal aging, resulting in lower circulating levels of IGF-I. The cause(s) of these deficits are complex and are enumerated in Table 1.[3–5] In healthy elderly people GH secretory pulse frequency is diminished, as are the GH secretory responses to administered GHRH and IGF-I production.[6–8] A diminished pituitary response to GHRH in aging rats appears to be related in part to deficiencies in the pituitary adenylate cyclase-cyclic adenosine monophosphate (cAMP) system.[9] Release of GHRH and GH after L-DOPA stimulation is diminished in elderly people, as are the GH responses to other

TABLE 1
Age-Related Changes in the GHRH-GH-IGF Axis

↓ central cholinergic tone leading to ↑ hypothalamic somatostatin
↓ hypothalamic GHRH mRNA and expression of pituitary GHRH receptors
↓ pituitary GH mRNA
↓ GHRH-induced GH secretion *in vivo* and *in vitro*
↓ GH secretory pulse frequency
↓ circulating GH
↓ serum GHBP and GH half-life
↓ IGF-I response to GH and to GHRH
↓ serum IGF-I and IGFBP-3 levels

traditional secretory stimuli.[10] *In vitro* correlates of these changes include a decreased abundance of pituitary GHRH receptors and of hypothalamic GHRH messenger RNA (mRNA) and pituitary GH mRNA.[4,11,12]

No central unifying basis has been shown conclusively to underlie these changes, but recent evidence suggests that depression of central cholinergic tone in the aging brain leads to enhanced somatostatinergic activity, which in turn depresses the somatotropic axis.[13,14] In this regard, interest has been considerable in the possibility that pharmacologic manipulation of the central nervous system may improve somatotropic function in older people. For example, Ceda and colleagues[15] tested the ability of the centrally acting drug, cytidine 5′-diphosphocholine (CDP), to alter GH secretion in healthy elderly volunteers. This drug alone induced a four-fold increase in serum GH concentrations. Adding GHRH to the infusion resulted in a GH response which was significantly greater than that observed after GHRH alone. Thus, CDP enhanced basal and GHRH-stimulated GH release in the elderly, although the mechanism of action of the drug remains unclear.

In addition to these central phenomena, plasma GH binding protein levels fall, and the half-life of endogenous GH is diminished in older subjects.[16,17] Serum IGF-I levels decline with age but still increase appropriately in response to GHRH or GH.[18,19] In women IGF binding protein (IGFBP) levels also change with aging, with a relative decrease in IGFBP-3.[20] To evaluate the capacity of older people to increase circulating IGF-I in response to GH, our group devised an IGF-generation test. Subjects receive 0.1 mg/kg of GH as a single subcutaneous injection and IGF-I is measured over the next 48 h. The levels of IGF-I peak at 24 to 32 h after the GH injection. In our experience with this test, older men had smaller increases in serum IGF-I than did young men. The responses of older, estrogen-depleted women did not differ from those of young women, although women who were taking stable doses of oral replacement estrogen had lower baseline levels of IGF-I and significantly diminished responses to GH.[21]

These results are of interest with regard to previous work that demonstrates lower mean 24 h GH and serum IGF-I levels in postmenopausal women than in premenopausal controls. The effects of estrogen in this system appear to depend greatly on the means by which it is given. In response to oral estrogen serum GH concentrations increase, but IGF-I levels decrease.[22] With transdermal estrogen, GH is apparently unchanged, but serum IGF-I either increases or remains stable.[23,24] These effects contrast with those of androgen, which appears not to be an important determinant of GH secretion in elderly men.

B. BODY COMPOSITION

That bone and lean body mass decrease and adiposity increases with age has become an axiom of gerontology and geriatric medicine. To account for these changes, several theories have been advanced, based either on behavioral or hormonal underpinnings. The behavioral model contends that normal aging is accompanied by substantial reduction in physical activity, and therefore in daily energy output and caloric requirement. These reductions tend not to be accompanied by reduced energy intake, leading to maintenance of body weight by increasing adipose mass. In addition, reduction in bone mass by this model reflects a decline in steady-state daily mechanical loading of the skeleton. Because serum IGF-I levels correlate with aerobic capacity, a behaviorally determined decline in physical fitness with age might also contribute to somatotropic decline.[25,26]

Hormonal models have been based on declines in reproductive and adrenal hormones as well as on somatotropic failure. The changes in body composition following menopause, particularly with respect to bone mass, are well described. Recent interest has focused on a role for dehydroepiandrosterone, a minor adrenal androgen.[27] Rudman[1] first proposed that some of the somatic features of normal aging, including osteopenia, muscle atrophy, frailty, and disordered sleep, may be caused in part by the decreased action of the GH-IGF-I axis. One piece of evidence that points toward such an effect can be seen in the features of younger adults with GH deficiency resulting from pituitary disease. Despite receiving adequate replacement therapy with thyroid, adrenal, and gonadal hormones, these patients have increased adiposity, particularly in visceral stores, decreased lean body mass, and osteopenia, all of which improve with rhGH replacement.[28]

Although the behavioral and hormonal models all have attractive elements, the basis for the somatic changes of aging are undoubtedly complex, and no single theory is likely ultimately to prove decisive. Nonetheless, the age-related decline in somatotropic function may contribute to a catabolic diathesis that eventuates in frailty, falls, and fractures in elderly people, a syndrome complex which has been named "the somatopause".[29]

C. EXERCISE AND BODY COMPOSITION IN THE ELDERLY

Although of some intrinsic interest, changes in body composition achieve clinical importance only when they are accompanied by altered body function. In the case of the reduced lean mass that attends GH deficiency evidence suggests that this deficit primarily represents loss of skeletal muscle and is associated with decreased muscular strength and endurance, which are themselves predictors of bone mineral density and of the risk of falling in older people.[30–33] Thus, muscle strength is likely to be an important determinant of hip fracture risk. Of the primary risk factors for fracture (muscle strength, neuromuscular coordination, visual acuity, and bone mass), it is probably easiest to enhance muscle strength in older persons so that it may prove a more effective strategy for diminishing hip fracture risk to develop means to increase muscle strength than to increase bone mineral density.[34]

It has been known for many years that older individuals gain muscle strength with exercise training, but until recently it was considered likely that increased strength reflected improvements in neurological factors rather than true muscle fiber hypertrophy.[35,36] We and others have shown that elderly men and women retain the capacity for muscle fiber hypertrophy with resistance training.[37–39] However, when we trained elderly subjects for up to 1 year, we found that strength gains rapidly plateaued, with very little additional strength improvement after 15

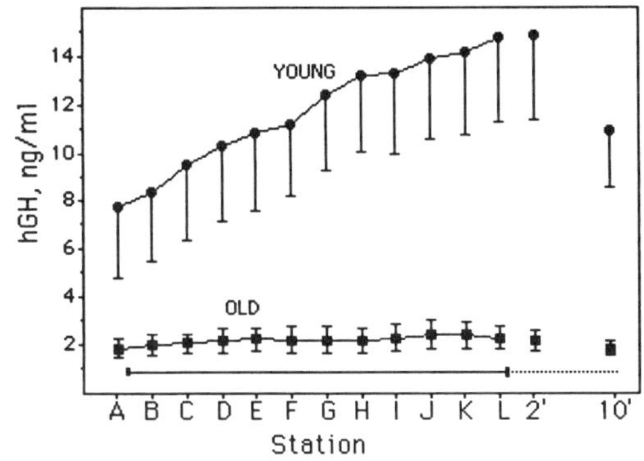

FIGURE 1. GH response to resistance exercise. Letters on horizontal axis correspond to baseline (A) and a series of 12 sequential resistance exercises (B to L). Two and 10 min values following L indicate minutes of recovery from exercise. Blood was drawn on completion of each activity. Data are given as mean ± SEM. (From Pyka et al., *J. Clin. Endocrinol. Metab.*, 75, 404, 1992. With permission.)

weeks despite maintaining the progressive nature of training.[40] Furthermore, the hypertrophy response in older subjects is still of relatively small magnitude compared to equivalent results in younger individuals exposed to a comparable training stimulus.[41]

Some evidence indicates a possible role for GH in the response to resistance training, and that GH deficiency may contribute to this age differential in hypertrophy response. Growth hormone, via IGF-I, stimulates muscle protein synthesis in normal humans *in vivo,* and IGF-I induces myofiber hypertrophy *in vitro.*[42,43] Both aerobic and resistance exercise acutely stimulates GH secretion, although the nature of GH response differs between the two types of activity.[44–48] With aerobic exercise, phasic increases in GH occur with the recovery phase rather than during the activity itself, whereas with resistance activity GH concentrations rise quickly during the exercise bout.[49,50] We recently examined the effect of age on the acute GH responses to resistance exercise.[48] In young subjects circulating GH rose within a few minutes after initiating a circuit of weightlifting activities. Growth hormone concentrations increased and remained elevated throughout the exercise period and decreased toward baseline during recovery. In the elderly subjects baseline GH values rarely increased beyond 6 ng/ml (Figure 1). Furthermore, training of elderly men and women for as much as 1 year achieved no systematic improvement in this GH secretory response. One plausible interpretation of these results, therefore, is that strength and hypertrophy responses of elderly men and women might be constrained by somatotropic failure. If this formulation is correct, treatment with GH during the course of an exercise program should result in enhanced strength and hypertrophy responses.

To test this possibility, we enrolled 20 elderly men (>65 years) into a 14-week program of progressive resistance exercise. All subjects exercised under supervision three times each week, and each exercise was carried out for three sets of eight repetitions at intensities of 65 to 70% of maximal strength. Subjects rapidly gained strength in all muscle groups, but at the end of 12 weeks all had reached a plateau, and no further strength was gained through week 14. At this point, the men were randomly assigned to receive rhGH, 0.25 mg/kg, or placebo as daily injections, and the exercise program and injections were carried out for an additional 10 weeks. Recombinant human GH had no effect on muscle strength or endurance for any muscle group, and in fact, increases in strength were not observed in either group between weeks 14 and 24 of the study. The only positive finding related to rhGH in this study was the observation that rhGH-treated men experienced a reduction in adipose mass and relative adiposity, and that this decrease was primarily localized to central adipose stores as determined by dual energy X-ray absorptiometry. This negative experiment suggests that limitations in muscle strength acquisition in the elderly cannot be ascribed simply to deficient somatotropic function.

IV. USE OF GROWTH HORMONE IN OLDER MEN AND WOMEN

Several groups have begun to systematically explore the effects of rhGH in older men and women, and the next section of this chapter summarizes this experience.

A. SHORT-TERM ADMINISTRATION OF RECOMBINANT HUMAN GROWTH HORMONE TO HEALTHY ELDERS

Our research group reported the effects of 7 d of rhGH administration to 16 healthy older men and women (>60 years).[19] Doses of rhGH ranged from 0.03 to 0.12 mg/kg/d, given as a single daily injection. The hormone produced a brisk, dose-dependent rise in circulating IGF-I, which appeared to reach steady state by about day 5. Treatment was associated with striking conservation of urinary nitrogen and sodium, but increased excretion of calcium. Thus, rhGH uncoupled the usual tight relationship between sodium and calcium excretion. Significant increases were observed in circulating osteocalcin and in urinary hydroxyproline, suggesting that bone remodeling had been activated. Of concern was the observation that insulin sensitivity, as determined during standard oral glucose tolerance tests, was impaired during the week of rhGH administration.

To learn how GH interacts with the parathyroid-vitamin D axis, we followed this study with an evaluation of the effects of GH on bone turnover and the response to parathyroid hormone (PTH) infusion in eight healthy elderly women.[64] Each subject was studied at baseline on short-term GH (0.06 mg/kg/d $\times$ 7 d) and on prolonged GH (0.02 mg/kg/d $\times$ 5 weeks). Under each condition, serum total and ionized calcium, phosphorus, osteocalcin, and urinary hydroxyproline were measured before and after an infusion of the synthetic peptide $PTH_{(1-34)}$. With short-term rhGH, basal IGF-I levels rose from 132 $\pm$ 13 to 553 $\pm$ 83 ng/ml. Basal calcium excretion rose 37%, and basal calcitriol levels rose 42%. Short-term rhGH did not alter plasma ionized calcium or inorganic phosphorus, the renal threshold for phosphorus excretion (TmP/GFR), circulating PTH, or any marker of bone turnover. $PTH_{(1-34)}$ produced an acute fall in TmP/GFR, a rise in urinary hydroxyproline excretion and serum calcitriol, and no change in serum osteocalcin. Short-term GH also increased the TmP/GFR response to PTH from 99.2 ± 22.3 to 144.1 ± 15.0 U ($p = 0.0021$), but did not affect hydroxyproline or calcitriol responses to PTH. With 5 weeks of rhGH, IGF-I levels (389 ± 60 ng/ml), calcium excretion (0.267 ± 0.020 mg/dl), serum inorganic phosphorus concentrations (4.55 ± 0.16 mg/dl), and basal TmP/GFR (4.29 ± 0.24 mmol/l) increased. Recombinant human GH did not affect basal plasma ionized calcium, serum PTH, or urinary hydroxyproline. Once again, the acute response of TmP/GFR to PTH infusion was enhanced on rhGH, with no effect on the calcitriol response to PTH. The cAMP response to $PTH_{(1-34)}$ infusion was similar during all three study periods. We concluded that GH antagonizes basal PTH effects on calcium and phosphorus handling by the kidney, but that phosphaturic responses to $PTH_{(1-34)}$ are enhanced. Because the cAMP response to PTH was unaffected by GH, these effects must occur distal to adenylate cyclase.

In another study of the rhGH effects on mineral metabolism Brixen et al.[51] showed that several daily injections of rhGH in young adult men initiated a prompt elevation in circulating osteocalcin. The remarkable feature of this study, however, was that osteocalcin levels remained elevated for many weeks after just a few days of exposure to rhGH.

B. SUSTAINED TREATMENT OF ELDERLY MEN

The most widely publicized growth hormone trial to date was reported by Rudman and colleagues (see chapter by Rudman, this volume).[52] These authors conducted a randomized, placebo-controlled intervention trial in 21 elderly men. Growth hormone (0.03 mg/kg three times per week) was given for 6 months, and was found to produce significant increases in lean mass by ^{40}K analysis. Bone density was assessed at nine different sites by dual photon absorptiometry, and a 1.6% increase in lumbar spine mineral density was reported. No significant change in skinfold estimates of adiposity was observed.

Although the results of this experiment were provocative and interesting, concerns about several aspects of their interpretation must be raised. The ^{40}K data provide convincing evidence of a true increase in lean mass. However, the stated changes in adiposity did not achieve statistical significance, and skinfold thickness is an inadequate method to assess adiposity in the face of fluid retention. The changes in bone mass were marginally significant at best.

C. SUSTAINED TREATMENT OF ELDERLY WOMEN

Our group recently conducted a randomized placebo-controlled trial of rhGH in 27 healthy elderly women >65 years of age, of whom eight took a stable dose of replacement estrogen. Growth hormone, at an initial daily dose of 0.043 mg/kg body weight (n = 19), or placebo (n = 14), was given as a single daily injection. After several weeks, 50% dose reductions were necessitated by side effects, primarily arthralgias, fluid retention, and, in two cases, carpal tunnel syndrome. The last seven subjects to be enrolled began treatment at this reduced level. A total of 13 women assigned to rhGH and 14 women assigned to placebo completed 6 months of drug treatment. In the rhGH group six women took oral estrogen replacement therapy (ERT), thus enabling separate analysis of the effects of rhGH according to estrogen status. In the rhGH group only, circulating levels of IGF-I and IGFBP-3 increased and stayed in the young adult range throughout the trial. However, women taking ERT had a significantly blunted response. Recombinant human GH led to a 9% increase in endogenous creatinine clearance. Moreover, in these women, rhGH did not alter a variety of cardiovascular risk factors, including carbohydrate tolerance, insulin sensitivity, lipoproteins, thyroid hormones, blood pressure, and circulating fibrinogen. In estrogen-deplete women rhGH significantly decreased LDL cholesterol levels and led to a transient increase in circulating triiodothyronine. In addition, a transient increase in plasma glucose and insulin was observed during an oral glucose tolerance test after 3 weeks of treatment. However, these had resolved to baseline levels by 6 weeks, where they remained throughout the balance of the trial. No differences were seen in nitrogen balance at 6 months. Body composition analysis by skinfold thickness, but not by hydrostatic weight, showed an apparent decrease in adiposity; no change was seen in lean mass by either technique.

Recombinant human GH led to striking increases in markers of bone turnover, particularly in the estrogen-deplete subjects. Sustained elevations were observed in urinary excretion of hydroxyproline and pyridinolines, as well as in circulating osteocalcin. No changes were observed in the type I procollagen extension peptide, another measure of bone formation activity. On the other hand, no significant changes were observed in bone mineral density at either the lumbar spine or proximal femur. It is of interest that bone mineral density at the femoral trochanter and Ward's triangle decreased significantly in the placebo group but did not change in the hormone-treated subjects. Thus, although rhGH did not increase bone mass, a maintenance effect at the hip apparently occurred. All changes in response to rhGH reverted to baseline values by 3 months after stopping treatment.

Similar to the results of our short-term trial, rhGH produced a substantial degree of fluid retention. At the lower dose of hormone used, this effect caused only mild symptoms of fluid retention in a few subjects. However, related changes in fluid balance may have produced alterations in total and extracellular water compartments that may confound the body composition measurements. For example, we observed the same changes in skin thickness that were reported by Rudman et al. within 7 days of starting rhGH injections.[52] At this time interval such a result must reflect water changes rather than a true loss of fat mass. This issue has not been adequately addressed and must be resolved in future studies.

V. A POTENTIAL ROLE FOR GROWTH HORMONE IN OSTEOPOROSIS THERAPY

Traditional models of GH action on the skeleton were predicated on an intermediary role for circulating IGFs, but recent evidence indicating that GH directly interacts with osteoblasts forced a reconsideration of this view. Growth hormone activates cell surface receptors in cultured osteoblast-like cells to increase IGF-I production.[53–55] In murine osteoblasts GH induces expression of c-*jun* and c-*fos* by a mechanism that involves the protein kinase C transduction complex.[56] In osteoblasts raised in serum-free medium GH stimulates proliferation and increases the relative amount of newly synthesized type I collagen, an effect which is obliterated by anti-IGF-I immunoglobulin.[57] Redundancy is apparent in the control of IGF-I production in osteoblasts because PTH and estradiol also promote accumulation of this somatomedin. *In vivo* GH and IGF-I both increase bone turnover.

In a classic experiment Harris and Heaney[58] showed that administration of GH to adult dogs increased bone mass. More recently rhGH was shown to maintain trabecular bone mass in primates rendered hypogonadal by a gonadotropin releasing hormone analogue.[59] Such *in vivo* demonstrations, combined with an increasing volume of *in vitro* evidence, invite the conclusion that GH or IGF-I may have the potential to activate osteoblast proliferation and differentiated function and to repair bone mineral deficits of older people in general, and of osteoporosis, in particular.

In humans significant relationships were observed between bone mineral density and circulating IGF-I, and men with idiopathic osteoporosis have been reported to have low IGF-I concentrations.[60,61]

The effects of GH described above suggest a rationale for considering the use of this hormone to increase bone mass in patients with osteopenia. Aloia et al.[62] reported an experience with GH in several osteoporotic patients, but the results did not support the clinical utility of this agent. In a subsequent report from this same group combined sequential therapy with GH and calcitonin produced a small, but persistent increase in whole body calcium content of osteoporotic patients.[63] Unfortunately, neither the results of Rudman et al. or our own experience (outlined above) permit enthusiasm that rhGH, given as single daily monotherapy, will offer a means to achieve meaningful increases in bone mass.[52] Although a maintenance effect may have occurred at the hip, a variety of antiresorptive agents currently offer similar protection, and it would be hard to justify the use of an expensive, injectable protein hormone to achieve this result. Because the doses employed are close to maximally tolerated levels, it is unlikely that a simple adjustment of dose will rectify this situation.

The weight of evidence clearly suggests that rhGH induces a prompt and sustained increase in bone remodeling. As bone remodeling is an inherently inefficient process, anything that activates remodeling might actually be predicted to aggravate bone loss. If a role exists for GH in osteoporosis, it will likely involve a complex therapeutic strategy. For example, it may be necessary to use rhGH cyclically in combination with an antiresorptive agent, so that a favorable shift in remodeling dynamics may be achieved.

Finally, it should be remembered that GH, at least transiently, promotes nitrogen retention in adults. Because decreased muscle mass and strength contribute to the risk for falls, and therefore to the risk for hip fracture, a therapeutic role for rhGH in osteoporosis may be defined that reflects improvement in muscle strength or endurance, rather than on bone mass per se.

VI. CONCLUSION

As the activity of the GH-IGF-I axis declines with aging, elderly individuals suffer from a constellation of events which has been termed the ''somatopause''. These include decreases in lean body and bone mass, increased adiposity, and a decline in renal function. It appears that rhGH replacement in older men and women may attenuate some aspects of these ''inevitable'' sequelae of aging. However, reports are conflicting and interpretation is complicated by the fact that intra and extracellular fluid shifts with rhGH may confound interpretation of the available studies. Perhaps most important, long-term therapy with rhGH has been accompanied by a high prevalence of important side effects, in particular, carpal tunnel syndrome. For future studies, although important changes in traditional cardiovascular risk factors appear not to have been produced by rhGH treatment, it will be important to maintain adequate surveillance over these and other potential risk factors, such as lipoprotein (a). In addition, future studies will investigate whether the desired effects of GH can be accomplished through the use of its mediator, IGF-I, a hormone which lacks some of the complicating side effects of GH itself.

REFERENCES

1. **Rudman, D.,** Growth hormone, body composition, and aging, *J. Am. Geriatr. Soc.*, 33, 800, 1985.
2. **Rosenfeld, R. G., Ocrant, I., Valentino, K. L., and Hoffman, A. R.,** Interaction of IGF with the hypothalamus and pituitary, in *Molecular and Cellular Biology of Insulin-Like Growth Factors and Their Receptors,* LeRoith, D. and Raizada, M. K., Eds., Plenum Press, New York, 1990, 39.
3. **Rudman, D., Kutner, M. H., Rogers, C. M., Lubin, M. F., Fleming, G. A., and Bain, R. P.,** Impaired GH secretion in the adult population: relation to age and adiposity, *J. Clin. Invest.*, 67, 1361, 1981.
4. **Hoffman, A. R., Griffin, C., Kalinyak, J. E., Perkins, S. N., and Ceda, G. P.,** The hypothalamic-somatotroph-somatomed in axis and aging, in *Psychoneuroendocrinology of Aging: Basic and Clinical Aspects,* Fidia Research Series, Vol. 16, Valenti, G., Ed., Springer-Verlag, Berlin, 1988, 43.
5. **Ceda, G., Denti, L., Hoffman, A. R., Ceresini, G., and Valenti, G.,** Aging and pituitary responses to hypothalamic peptides, in *Protein Metabolism in Aging,* Segal, H. L., Ed., Wiley-Liss, New York, 1990, 335.

6. **Ho, K. Y., Evans, W. S., Blizzard, R. M., Veldhuis, J. D., Merriam, G. R., Samojlik, E., Furlanetto, R., Kaiser, D. L., and Thorner, M. O.,** Effects of sex and age on the 24-hour profile of growth hormone secretion in man: importance of endogenous estradiol concentrations, *J. Clin. Endocrinol. Metab.*, 64, 51, 1987.
7. **Shibasaki, T., Shizume, K., Nakahara, M., Masuda, A., Jibiki, K., Demura, H., Wakabayashi, T., and Ling, N.,** Age-related changes in plasma growth hormone response to growth hormone releasing factor in man, *J. Clin. Endocrinol. Metab.*, 58, 212, 1984.
8. **Pavlov, E. P., Harman, S. M., Merriam, G. R., Gelato, M. C., and Blackman, M. R.,** Responses of growth hormone (GH) and somatomedin-C to GH-releasing hormone in healthy aging men, *J. Clin. Endocrinol. Metab.*, 62, 595, 1986.
9. **Ceda, G. P., Valenti, G., Butturini, U., and Hoffman, A. R.,** Diminished pituitary responsiveness to growth hormone releasing factor in aging male rats, *Endocrinology*, 118, 2109, 1986.
10. **Bando, H., Zhang, C., Takada, Y., Yamasaki, R., and Saito, S.,** Impaired secretion of growth hormone-releasing hormone, growth hormone and IGF-I in elderly men, *Acta Endocrinol.*, 124, 31, 1991.
11. **Abribat, T., Deslauriers, N., Brazeau, P., and Gaudreau, P.,** Alterations of pituitary GHRH binding sites in aging rats, *Endocrinology*, 128, 633, 1991.
12. **De Gennaro Colonna, V., Zoli, M., Cocchi, D., Maggi, A., Marrama, P., Agnati, L. F., and Muller, E. E.,** Reduced growth hormone-releasing factor (GHRF) - like immunoreactivity and GHRF gene expression in hypothalamus of aged rats, *Peptides*, 10, 705, 1989.
13. **Pepeu, G., Casamenti, F., Pepeu, I. M., and Scali, C.,** The brain cholinergic systemi in ageing mammals, *J. Reprod. Fertil.*, 46(Suppl.), 155, 1993.
14. **Müller, E. E., Cella, S. G., De Gennaro Colonna, V., Parenti, M., Cocchi, D., and Locatelli, V.,** Aspects of the neuroendocrine control of growth hormone secretion in ageing mammals, *J. Reprod. Fertil.*, 46(Suppl.), 99, 1993.
15. **Ceda, G. P., Ceresini, G., Denti, L., Magnani, D., Marchini, L., Valenti, G., and Hoffman, A. R.,** Effects of cytidine 5′-diphosphocholine administration on basal and growth hormone-releasing hormone-induced growth hormone secretion in elderly subjects, *Acta Endocrinol.*, 124, 516, 1991.
16. **Daughaday, W. H., Trivedi, B., and Andrews, B. A.,** The ontogeny of serum GH binding protein in man: a possible indicator of hepatic GH receptor development, *J. Clin. Endocrinol. Metab.*, 65, 1072, 1987.
17. **Iranmanesh, A., Lizarralde, G., and Veldhuis, J. D.,** Age and relative adiposity are specific negative determinants of the frequency and amplitude of growth hormone (GH) secretory bursts and the half-life of endogenous GH in elderly men, *J. Clin. Endocrinol. Metab.*, 73, 1081, 1991.
18. **Corpas, E., Harman, S. M., Pineyro, M. A., Roberson, R., and Blackman, M. R.,** Growth hormone (GH)-releasing hormone-(1–29) twice daily reverses the decreased GH and insulin-like growth factor-I levels in old men, *J. Clin. Endocrinol. Metab.*, 75, 530, 1987.
19. **Marcus, R., Butterfield, G., Holloway, L., Gilliland, L., Baylink, D. J., Hintz, R. L., and Sherman, B. L.,** Effects of short-term administration of recombinant human growth hormone to elderly people, *J. Clin. Endocrinol. Metab.*, 70, 519, 1990.
20. **Donahue, L. R., Hunter, S. J., Sherblom, A. P., and Rosen, C.,** Age-related changes in serum insulin-like growth factor-binding proteins in women, *J. Clin. Endocrinol. Metab.*, 71, 575, 1990.
21. **Lieberman, S. A., Mitchell, A. M., Marcus, R., Hintz, R. L., and Hoffman, A. R.,** The insulin-like growth factor I generation test: resistance to growth hormone with aging and estrogen replacement therapy, *Horm. Metab. Res.*, 26, 229–233, 1994.
22. **Dawson-Hughes, B., Stern, D., Goldman, J., and Reichlin, S.,** Regulation of growth hormone and somatomedin-C secretion in postmenopausal women: effect of physiological estrogen replacement, *J. Clin. Endocrinol. Metab.*, 63, 424, 1986.

23. **Weissberger, A. J., Ho, K. K. Y., and Lazarus, L.,** Contrasting effects of oral and transdermal routes of estrogen replacement therapy on 24-hour growth hormone (GH) secretion, insulin-like growth factor I, and GH-binding protein in postmenopausal women, *J. Clin. Endocrinol. Metab.,* 72, 374, 1991.
24. **Bellantoni, M. F., Harman, S. M., Cho, D. E., and Blackman, M. R.,** Effects of progestin-opposed transdermal estrogen administration on growth hormone and insulin-like growth factor-I in postmenopausal women of different ages, *J. Clin. Endocrinol. Metab.,* 72, 172, 1991.
25. **Kelly, P. J., Eisman, J. A., Stuart, M. C., Pocock, N. A., Sambrook, P. N., and Gwinn, T. H.,** Somatomedin-C, physical fitness, and bone density, *J. Clin. Endocrinol. Metab.,* 70, 718, 1990.
26. **Poehlman, E. T., and Copeland, K. C.,** Influence of physical activity on insulin-like growth factor-I in healthy younger and older men, *J. Clin. Endocrinol. Metab.,* 71, 1468, 1990.
27. **Rannevik, G., Carlstrom, K., Jeppsson, S., Bjerre, B., and Svanberg, L.,** A prospective long-term study in women from pre-menopause to post-menopause: changing profiles of gonadotrophins, oestrogens and androgens, *Maturitas,* 8, 297, 1986.
28. **Bengtsson, B.-A., Edén, S., Lönn, L., Kvist, H., Stokland, A., Lindstedt, G., Bosaeus, I., Tölli, J., Sjöström, L., and Isaksson, O. G. P.,** Treatment of adults with growth hormone (GH) deficiency with recombinant human GH, *J. Clin. Endocrinol. Metab.,* 46, 309, 1993.
29. **Hoffman, A. R., Pyka, G., Lieberman, S. A., Ceda, G. P., and Marcus, R.,** The somatopause, in *Growth Hormone and Somatomedins During Lifespan,* Muller, E. E., Cocchi, D., and Locatelli, V., Eds., Springer-Verlag, Berlin, 1993, 265.
30. **Bevier, W. C., Wiswell, R. A., Pyka, G., Kozak K. C., Newhall, K. M., and Marcus, R.,** Relationship of body composition, muscle strength, and aerobic capacity to bone mineral density in older men and women, *J. Bone Min. Res.,* 4, 421, 1989.
31. **Snow-Harter, C., Bouxsein, M., Lewis, B., Charette, S., Weinstein, P., and Marcus, R.,** Muscle strength as a predictor of bone mineral density in young women, *J. Bone Min. Res.,* 5, 589, 1990.
32. **Wolfson, L. I., Whipple, R., Amerman, P., Kaplan, J., and Kleinberg, A.,** Gait and balance in the elderly: two functional capacities which link sensory and motor ability to falls, *Clin. Geriat. Med.,* 1, 649, 1985.
33. **Whipple, R. H., Wolfson, L. I., and Amerman, P. M.,** The relationship of knee and ankle weakness to falls in nursing home residents: an isokinetic study, *J. Am. Geriatr. Soc.,* 35, 13, 1987.
34. **Hayes, W. C., Piazza, S. J., and Zysset, P. K.,** Biomechanics of fracture risk prediction of the hip and spine by quantitative computed tomography, *Radiol. Clin. N. Am.,* 29, 1, 1991.
35. **Moritani, T., and deVries, H. A.,** Neural factors versus hypertrophy in the time course of muscle strength gain, *Am. J. Phys. Med.,* 58, 115, 1979.
36. **Moritani, T., and deVries H. A.,** Potential for gross muscle hypertrophy in older men, *J. Gerontol.,* 35, 672, 1980.
37. **Charette, S., McEvoy, L., Pyka, G., Snow-Harter, C., Guido, D., Wiswell, R. A., and Marcus, R.,** Muscle hypertrophy response to resistance training in older women, *J. Appl. Physiol.,* 70, 1912, 1991.
38. **Frontera, W. R., Meredith, C. N., O'Reilly, K. P., Knuttgen, H. G., and Evans, W. J.,** Strength conditioning in older men: skeletal muscle hypertrophy and improved function, *J. Appl. Physiol.,* 64, 1038, 1988.
39. **Fiatarone, M. A., Marks, E. C., Ryan, N. D., Meredith, C. N., Lipsitz, L. A., and Evans, W. J.,** High-intensity strength training in nonagenarians, *JAMA,* 263, 3029, 1990.
40. **Pyka, G., Lindenberger, E., Charette, S., and Marcus, R.,** Muscle strength and fiber adaptations to a year-long resistance training program in elderly men and women, *J. Gerontol.,* 49, M22–M27, 1994.
41. **Staron, R. S., Malicky, E. S., Leonardi, M. J., Falkel, J. E., Hagerman, F. C., and Dudley, G. A.,** Muscle hypertrophy and fast fiber type conversions in heavy resistance-trained women, *Eur. J. Appl. Physiol.,* 60, 71, 1989.

42. **Fryberg, D. A., Gelfand, R. A., and Barrett, E. J.,** Growth hormone acutely stimulated forearm muscle protein synthesis in normal humans, *Am. J. Physiol.*, 260, 499, 1991.
43. **Vandenburgh, H. H., Karlisch, P., Shewsky, J., and Feldstein, R.,** Insulin and IGF-1 induce pronounced hypertrophy of skeletal muscle myofibers in tissue culture, *Am. J. Physiol.*, 260, 475, 1991.
44. **Lassarre, C., Girard, F., Durand, J., and Raynaud, J.,** Kinetics of human growth hormone during submaximal exercise, *J. Appl. Physiol.* 37, 826, 1974.
45. **Nilsson, K. O., Heding, L. G., and Hokfelt, B.,** The influence of short term submaximal work on the plasma concentrations of catecholamines, glucagon and GH in man, *Acta Endocrinol.*, 79, 286, 1975.
46. **Sutton, J. R.,** Hormonal and metabolic responses to exercise in subjects of high and low work capacities, *Med. Sci. Sports Exercise,* 10, 1, 1978.
47. **Hagberg, J. M., Seals, D. R., Yerg, J. E., Gavin, J., Gingerich, R., and Premachandra, R.,** Metabolic responses to exercise in young and older athletes and sedentary men, *J. Appl. Physiol.*, 65, 900, 1988.
48. **Pyka, G., Wiswell, R. A., and Marcus, R.,** Age-dependent effect of resistance exercise on growth hormone secretion in people, *J.Clin. Endocrinol. Metab.*, 75, 404, 1992.
49. **Hartley, C. H., Mason, J. W., Mogan, R. P., Jones, L. G., Kotchen, T. A., Mougey, E. H., Wherry, F. E., Pennington, L. L., and Ricketts, P. T.,** Multiple hormone responses to graded exercise in relation to training, *J. Appl. Physiol.*, 33, 602, 1972.
50. **Vanhelder, W. P., Goode, R. C., and Radomski, M. W.,** Effect of anaerobic and aerobic exercise of equal duration and work expenditure on plasma GH levels, *Eur. J. Appl. Physiol.*, 52, 255, 1984.
51. **Brixen, K., Nielsen, H. K., Mosekilde, L., and Flyvbjerg, A.,** A short course of recombinant human growth hormone treatment stimulates osteoblasts and activates bone remodeling in normal human volunteers, *J. Bone Min. Res.*, 5, 609, 1990.
52. **Rudman, D., Feller, A. G., Nagraj, H. S., Gergans, G. A., Lalitha, P. Y., Goldberg, A. F., Schlenker, R. A., Cohn, L., Rudman, I. W., and Mattson, D. E.** Effects of human growth hormone in men over 60 years old, *N. Engl. J. Med.*, 323, 1, 1990.
53. **Stracke, H., Schultz, A., Moeller, D., Rossol, S., and Schatz, H.,** Effect of growth hormone on osteoblasts and demonstration of somatomedin C/IGF-1 in bone organ culture, *Acta Endocrinol. (Copenhagen),* 107, 16, 1984.
54. **Chenu, C., Valentin-Opran, A., Chavassieux, P., Saez, S., Meunier, P. J., and Delmas, P. D.,** Insulin like growth factor I hormonal regulation by growth hormone and by $1,25(OH)_2D_3$ and activity on human osteoblast-like cells in short-term cultures, *Bone,* 11, 81, 1990.
55. **Barnard, R., Ng, K. W., Martin, T. J., and Waters, M. J.,** Growth hormone (GH) receptors in clonal osteoblast-like cells mediate a mitogenic response to GH, *Endocrinology,* 128, 1459, 1991.
56. **Slootweg, M. C., de Groot, R. P., Herrmann-Erlee, M. P. M., Koornneef, I., Kruijer, W., and Kramer, Y. M.,** Growth hormone induces expression of c-jun and jun B oncogenes and employs a protein kinase C signal transduction pathway for the induction of c-fos oncogene expression, *J. Mol. Endocrinol.*, 6, 179, 1991.
57. **Ernst, M., and Froesch, E. R.,** Growth hormone dependent stimulation of osteoblast-like cells in serum-free cultures via local synthesis of insulin-like growth factor I, *Biochem. Biophys. Res. Commun.*, 151, 142, 1988.
58. **Harris, W. H., and Heaney,** R. P. Effect of growth hormone on skeletal mass in adult dogs, *Nature,* 273, 403, 1969.
59. **Mann, D. R., Rudman, C. G., Akinbami, M. A., and Gould, K. G.,** Preservation of bone mass in hypogonadal female monkeys with recombinant human growth hormone administration, *J. Clin. Endocrinol. Metab.*, 74, 1263, 1992.
60. **Johansson, A. G., Burman, P., Westermark, K., and Ljunghall, S.,** The bone mineral density in acquired growth hormone deficiency correlates with circulating levels of insulin-like growth factor I, *J. Int. Med.*, 232, 447, 1992.

61. **Ljunghall, S., Johansson, A. G., Burman, P., Lindh, E., and Karlsson, F. A.,** Low plasma levels of insulin-like growth factor 1 (IGF-1) in male patients with idiopathic osteoporosis, *J. Int. Med.,* 232, 59, 1992.
62. **Aloia, J. F., Zanzi, I., Ellis, K., and Jowsey, J.,** Effects of growth hormone in osteoporosis, *J. Clin. Endocrinol. Metab.,* 43, 992, 1976.
63. **Aloia, J. F., Vaswani, A., Kapoor, A., Yeh, J. K., and Cohn, S. H.,** Treatment of osteoporosis with calcitonin, with and without growth hormone, *Metabolism,* 34, 124, 1985.
64. **Lieberman, S. A.** et al., 1993, unpublished data.

Chapter 9

EFFECTS OF HUMAN GROWTH HORMONE IN OLDER ADULTS

Adil A. Abbasi and Daniel Rudman

TABLE OF CONTENTS

I. EFFECTS OF AGING ON THE HYPOTHALAMUS-GROWTH HORMONE-INSULIN-LIKE GROWTH FACTOR-I AXIS

Growth hormone is a single-chain, 22,000 mol wt polypeptide with 191 amino acids which is produced by the somatotropic cells of the anterior pituitary gland. Its release from the anterior pituitary is regulated by GH releasing hormone (GHRH) and somatostatin (SRIH).[1–4] Hypothalamic GHRH and SRIH are released into the pituitary portal venous system under neuroregulatory control and exert opposing actions on the synthesis and release of GH. In human subjects GH secretion occurs in pulses, primarily in the slow wave or nonrapid eye movement sleep stages 3 and 4.[5,6] Peripheral actions of GH are mediated primarily by insulin-like growth factor-I (IGF-I),[7–9] which is produced at extrahepatic sites as well as in the liver under the influence of GH.[10] The liver, however, is the primary site for the production of circulating IGF-I.[11] Serum (plasma) IGF-I levels reflect GH secretion, and low IGF-I levels generally indicate low GH secretion.[12,13] In men and women the average serum IGF-I concentration declines progressively from the fourth through the ninth decades.[14,15,19] This decline in IGF-I with aging is associated with diminished pulsatile release of GH from the anterior pituitary gland.[16–18] Based on the evidence from both animal and human experiments, it is

0-8493-8384-6/95/$0.00+$.50

TABLE 1
Summary of the IGF-I Levels in the Various Groups of Subjects

	Healthy young men (n = 32)	Healthy old men (n = 27)	Institutionalized old men (n = 117)
Age (years)			
Average	25.1	70.0	72.9
Range	20–29	59–98	59–95
Body mass index (kg/m^2)			
Average	24.6	23.6	26.0
Range	18.9–34.4	20.2–27.3	13.6–40.5
IGF-I (ng/ml)			
Average	361[a]	170	154
SD	87.6	55.8	65.0
95% interval	240–460	80–290	67–316

[a] Different from the other groups at $p < 0.01$.
From Abbasi, A. A. et al., *J. Am. Geriatr. Soc.*, 41, 975, 1993. With permission.

believed that the decline in GH release from the aging pituitary gland is due to altered hypothalamic regulation involving both a decline in GHRH activity and an increase in SRIH activity.[1–4]

II. INSULIN-LIKE GROWTH FACTOR-I LEVELS IN THE ELDERLY

While the circulating levels of IGF-I tend to decline with advancing age in both men and women,[14–19] interindividual variation is considerable.[15] The authors recently published a survey comparing IGF-I levels among three groups of men:[15] healthy young men, healthy old men, and chronically institutionalized or infirm old men. A low IGF-I level (defined as a value below the lower 2.5 percentile of the comparison group) occurred in 80% of the healthy old men and in 90% of the institutionalized old men, when compared with the healthy young men ($p < 0.001$); and in 26% of the institutionalized old men when compared with the healthy old men ($p < 0.001$). In the healthy or infirm old men with low IGF-I levels as compared to healthy young men nocturnal peaks of serum GH were generally <2 ng/ml, indicating the central cause of hyposomatomedinemia. Table 1 shows the IGF-I levels from the three groups under consideration. The IGF-I level was highest in the healthy young men (mean 361 ng/ml). The mean values in the healthy old men (mean 170 ng/ml) and in the infirm old men (mean 154 ng/ml) were significantly lower than the healthy young men ($p < 0.05$). By the Kolmogorov-Smirnov test of the distribution curves, there differences were significant ($p < 0.05$) between all three groups (Figure 1). Table 1 shows that for healthy young men, the 95% reference range for IGF-I is 240 to 460 ng/ml. Values <240 ng/ml are below the lower 2.5 percentile of the healthy young men and

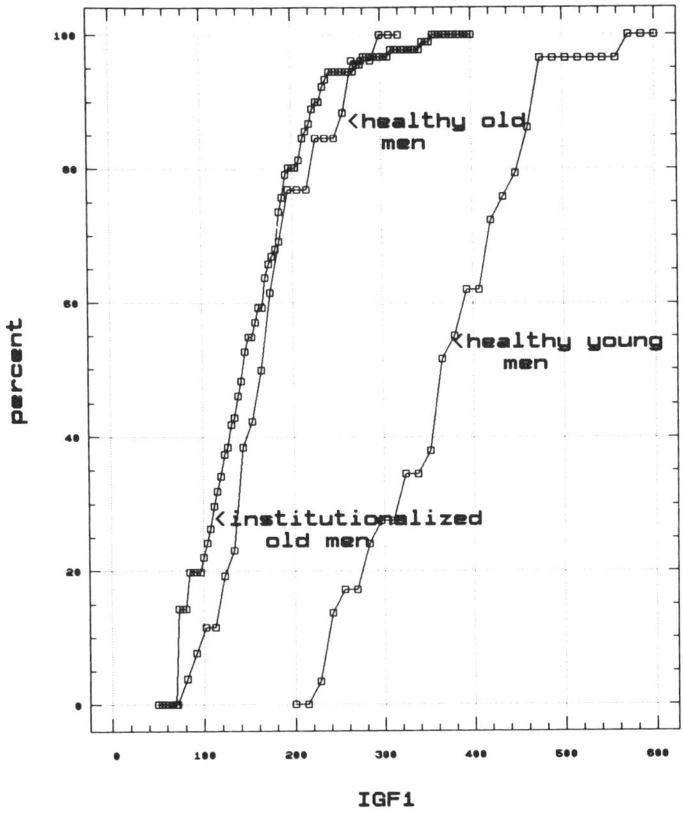

FIGURE 1. Kolmogorov-Smirnov test of the distribution curves comparing IGF-I levels in three groups of men. (From Abbasi, A. A. et al., *J. Am. Geriatr. Soc.*, 41, 975, 1993. With permission.)

represent hyposomatomedinemia, employing healthy young men as the reference standard. Using this definition, 80% of the healthy old men and 90% of the infirm old men are hyposomatomedinemic (Figure 1). Several other studies showed similar results.[1,16] The serum IGF-I levels tend to be higher in younger females than males, but no such sex difference is observed in older subjects.[16,19]

III. GROWTH HORMONE REPLACEMENT IN ADULTS

Studies on GH replacement in adults can be divided into three categories: (1) adults age 20 to 50 years with a history of GH deficiency, associated either with growth failure in childhood or with a pituitary lesion acquired in adulthood; (2) adults age 60 to 90 years with no history of GH deficiency, but with low IGF-I or GH levels as compared to their younger counterparts; and (3) adults of various ages with clinical problems such as malnutrition, recent trauma, or osteoporosis.

A. ADULTS WITH HISTORY OF GROWTH HORMONE DEFICIENCY

This group of adults includes subjects with isolated GH deficiency or multiple pituitary hormone deficits.[20–30] The latter conditions were treated with appropriate supplements of thyroxine, adrenal, and gonadal hormones.

Some of the recently published studies demonstrating physiologic and psychologic deficits in GH-deficient adults are summarized in Table 2. Growth hormone-deficient adults tend to have decreased lean body mass, muscle mass, and exercise performance.[23,27,28,30] They tend to have higher body fat mass.[23,28,30] This deficiency also seems to be associated with decreased creatinine clearance and renal plasma flow.[30] The adults have smaller left ventricular mass, smaller interventricular septum, and decreased left ventricular ejection fraction.[24,25] They also tend to have diminished cardiac performance with exercise.[24] These changes in cardiac function in GH-deficient adults are significantly related to the degree of GH deficiency and may be the cause of increased cardiovascular mortality in patients with treated hypopituitarism.[24] Growth hormone-deficient adults also have significant psychological impairment as predicted by "quality of life assessment".[22,29] Most of the above-mentioned adverse effects of GH deficiency can be reversed by replacement doses of GH.[20–23,27–30]

Administration of human GH (hGH) to this group of adults was shown to positively influence body composition, muscle performance, cardiac function, aerobic work capacity (VO_2 max), and psychometrics.[22,23,27–29] Administration of hGH to this group of adults results in a significant increase in the muscle:fat ratio and in the lean body mass.[23,28,30] Muscle strength, aerobic work capacity, endurance, as well as increased basal- and exercise-induced cardiac output were observed after short-term treatment of this group of adults with GH.[20,21,23,26–28] Improvement in mood and behavior with fewer psychosomatic complaints was also noted following GH replacement.[22,29] Other changes noted following GH replacement in this group included decline in serum cholesterol level, fasting hyperglycemia, rise in serum calcium and osteocalcin level, as well as improvement in the glomerular filtration rate, renal plasma flow, and creatinine clearance.[20,21,30]

Adverse effects of GH therapy in this group included weight gain, peripheral edema secondary to fluid retention, carpal tunnel syndrome, and joint pain.[22,24]

B. ADULTS WITH NO HISTORY OF GROWTH HORMONE DEFICIENCY, BUT LOW LEVELS OF INSULIN-LIKE GROWTH FACTOR-I OR GROWTH HORMONE

This group of adults included subjects with no history of GH deficiency, but found to have low IGF-I levels as compared to their younger counterparts.[31–35] Some of the changes that are observed with aging, e.g., decreased lean body mass, increased adipose mass, and decline in skeletal muscle mass and bone mass, are similar to changes observed in patients with GH deficiency.[36] Based on this comparison, it was postulated that some of the changes which take place with aging are secondary to low circulating levels of IGF-I. Studies done to test this hypothesis measured metabolic and body composition changes which take place after GH replacement in the elderly with either low IGF-I levels or inadequate GH secretion. Some of the recently published studies for this group of older adults are summarized in Table 3.

Marcus et al.[31] studied 18 subjects (12 females, 6 males) older than 60 years, with suboptimal nocturnal GH release but IGF-I level similar to a younger control group. This was a 7-d trial in which subjects were randomly assigned to receive recombinant human growth hormone (rhGH) at doses 0.03, 0.06, or 0.12 mg/kg/day subcutaneously for a period of 7 d. The administration of GH was associated with an increase in the serum concentration of osteocalcin, calcitriol, parathyroid hormone, and urinary hydroxyproline, and a positive nitrogen balance. The authors reported edema, glucose intolerance, and hyperinsulinism as adverse effects of GH administration.

Rudman et al.[32] studied 45 independent men over 60 years of age with plasma IGF-I levels below 0.35 U/ml, which is significantly lower than those of young healthy men. This was a 21-month study in which the first 6 months was the observation period, followed by a 12-month treatment period and a 3-month post-treatment observation period. During the treatment period, 26 men (group I) received 0.03 mg/kg of rhGH subcutaneously three times a week, and 19 men (group II) received no treatment. The following outcome variables were measured at 0, 6, 12, and 18 months: lean body mass (LBM); adipose mass (AM); skin thickness; size of the liver, spleen and kidneys; the cross-sectional areas of ten muscle groups; and bone density at nine skeletal sites. Lean body mass and AM were also measured at 21 months. In group I, hGH treatment raised the plasma IGF-I level in the range of 0.5 to 1.5U/ml. Significant changes occurred in the following measurable outcomes, expressed as percent change at 18 months over baseline as shown in Table 4: LBM +6%, AM −15%, skin thickness +4%, liver volume +8%, spleen volume +23%, sum of ten muscle areas +11%. Three months after hGH treatment was stopped, about one half of the hGH-induced increase in LBM disappeared and about one third of the hGH-induced decrease in AM reappeared. In group II (control group), the IGF-I level remained below 0.35 U/ml. At the 18-month period, the untreated controls showed a significant decline in LBM to 96% and skin thickness to 94% of initial baseline.

A trial with replacement doses of hGH in 83 healthy elderly men (mean age, 69 years) with plasma IGF-I level below 0.35 U/ml was done by Cohn et al.[33] The target range for the serum IGF-I level was 0.5 to 1.5 U/ml. Table 5 shows the relationship between adverse effects and IGF-I levels during hGH treatment in this trial. Subjects whose mean intratreatment IGF-I level was 0.5 to 1.0 U/ml responded with a 7% increase in LBM (anabolic effect) and an 18% decrease in adipose mass (lipolytic effect) and experienced no adverse drug reactions. Those men with average intratreatment IGF-I levels of 1.0 to 1.5 U/ml had lesser anabolic (+3%) and lipolytic effects (−12%), and a substantial incidence of adverse drug reactions. The adverse side effects disappeared spontaneously within 3 months

TABLE 2
Summary of Recently Published Studies Demonstrating Physiologic Deficits in Growth Hormone-Deficient Adults

Author	Study group	Conclusion
Shahi, M., et al.[24]	Twenty-six patients, aged 26–64 years (mean age, 46 years) with treated hypopituitarism but GH deficient, were studied by cross-sectional and doppler echocardiography and by exercise testing. The results were correlated with the degree of GH deficiency.	The left ventricular mass and maximum cardiac work achieved on exercise were significantly related to the degree of GH deficiency. There was a significant correlation between IGF-I and left ventricular mass (r = 0.45, p <0.02). Lean body mass was also significantly correlated with the left ventricular mass (r = 0.78, p <0.0001) and left ventricular diastolic function (r = −0.63, p <0.01). There was a significant correlation between serum IGF-I and the rate-pressure product on exercise (r = 0.47, p<0.01). Increased incidence of cardiovascular mortality in patients with treated hypopituitarism may be secondary to abnormalities of cardiac structure and function demonstrated in this study, which seems to be related to GH deficiency.
Merola, B., et al.[25]	Twenty patients, aged 21–33 years (mean age, 27 years) with the diagnosis of GH deficiency since childhood were studied. The cardiac function was evaluated using 1- and 2-dimensional echocardiography. All results were compared with age- and sex-matched normal controls.	GH-deficient patients had significantly lower values of interventricular septum (7.1 ± 1 vs. 9 ± 0.4 mm, p<0.01) and left ventricular posterior wall thickness (6.1 ±1 vs. 9 ± 0.04 mm, p <0.01), which resulted in a significantly smaller left ventricular mass index (54± 11 vs. 85 ± 15 g/m^2; p <0.001). Left ventricular ejection fraction was also lower in the GH-deficient patients (59± 9% vs. 69 ± 10% p <0.05).
Salomon et al.[30]	Twenty-four patients, aged 21–51 years, with the history of GH deficiency were treated with replacement doses of GH in a double-blind, placebo-controlled trial for a period of 6 months.	GH deficiency causes a decrease in lean body mass, an increase in body fat content, and a decrease in basal metabolic rate. In patients treated with GH, lean body mass and basal metabolic rate increased and body fat decreased (p <0.01). Fasting plasma glucose, insulin and c-peptide increased during treatment with growth hormone (p <0.01). Plasma cholesterol concentration decreased during treatment with GH (p <0.05). Creatinine clearance was lower in the GH deficient patients, but increased after treatment with GH (p <0.01). Authors concluded, based on their findings, that GH has a role in the regulation of body composition in adults.
Jørgensen et al.[23]	Twenty-two GH-deficient adults with a mean age of 23.8 years were treated with replacement doses of GH in a double-blind, placebo-controlled crossover study.	GH deficiency causes a decrease in muscle volume and an increase in adipose tissue. These changes in body composition are reversed following treatment with GH. The mean muscle volume of the thigh, estimated by computerized tomography, was significantly higher after GH than after placebo (70.0 vs. 66.3 ml/0.8 cm cross-sectional slice; p <0.01) ora 5.6% increase following treatment with GH.

		GH-deficient patients had lower exercise capacity which improved significantly after GH replacement (60.8 vs. 54.2 kJ; $p < 0.05$) or a 12% increase following treatment with GH.
Cuneo et al.[26]	Twenty-four GH-deficient adults, aged 21–51 years (mean age, 39 years) were studied to assess skeletal muscle mass and function and compared to healthy untrained controls.	Compared to healthy untrained controls, adults with GH deficiency had significantly reduced cross-sectional area of thigh muscle/body weight (1.58 vs. 1.87 $p < 0.01$), or 18% less than the untrained controls, reduced quadriceps force/weight (5.54 vs. 7.49; $p < 0.01$), or 35% less than the untrained controls and reduced quadriceps force/quadriceps area (6.91 vs. 7.89; $p < 0.05$), or 14% less than the untrained controls.
Cuneo et al.[27]	Twenty-four patients aged 21–51 years with history of growth GH deficiency were treated with replacement doses of GH in a double-blind, placebo-controlled trial for a period of 6 months, to assess the effects of GH replacement on exercise performance.	GH-deficient patients had lower exercise performance as measured by $VO_{2\,max}$ and maximal power output; both $VO_{2\,max}$ and maximal power output improved significantly (17 and 19%, respectively) following treatment with replacement doses of GH as compared to placebo. $VO_{2\,max}$ in the GH treatment group was +406 vs. 133 ml/min in the placebo group; $p < 0.01$ and maximal power output was +24.6 in the GH treatment group vs. +9.7 in the placebo group; $p < 0.05$.
Whitehead et al.[28]	Fourteen GH-deficient adults aged 19–52 years (mean age, 29 years) were treated with replacement doses of GH in a double-blind placebo-controlled crossover study.	Patients with GH deficiency had lower lean body mass and higher fat mass. Following treatment with GH lean body mass increased and fat mass decreased ($p < 0.05$). GH-deficient patients had lower thigh muscle volume and exercise capacity, both of which increased following treatment with GH. Thigh muscle volume increased following GH treatment to 99.5 vs. 94.1 ml/0.8mm computerized tomographic slice before GH treatment (or increased by 6%); $p < 0.05$. Exercise capacity increased from 174 to 199 W following GH treatment (or increased by 14%); $p < 0.05$. $VO_{2\,max}$ also increased from 1.93 to 2.17 l/m following treatment with GH (or increased by 12%); $p < 0.05$.
Degerblad et al.[22]	Six patients with history of GH deficiency, aged 21–50 years, were treated with replacement doses of GH.	All patients following treatment reported improved well being and increased work capacity. 83% of patients showed improvement in bone mineral density as measured by single photon absorptiometry.
McGauley et al.[29]	Twenty-four adults with GH deficiency, aged 18–55 years, were treated with the replacement doses of GH in a double-blind, placebo-controlled trial for a period of 6 months.	Significantly lower scores on ''quality of life'' assessment in the GH-deficient patients were reported prior to treatment with GH as compared to the matched controls. Following replacement with GH, significant psychological improvement in energy level and mood was recorded as compared to the placebo group.

TABLE 3
Summary of Clinical Trials Using hGH in the Elderly

Author	Study group	Design	GH regimen	Effects	Adverse effects
Marcus et al.[31]	12 healthy women and 6 healthy men (mean age, 67.7 years).	After 10 d of equilibration rhGH was administered for 7 d.	Subjects were randomly assigned to receive 0.03, 0.06, or 0.12mg/kg of rhGH injections sc for 7 d.	Decrease in urinary nitrogen and sodium excretion. Increase in serum levels of phosphate, osteocalcin, calcitriol, and PTH. Fasting plasma cholesterol declined.	All subjects on 0.12 mg/kg dose of rhGH developed glucose intolerance.
Rudman et al.[32]	45 healthy men with IGF-I level <0.35 U/ml (age, 61–81 years).	Two groups were studied. Treatment group (I) (26 men) was given rhGH and control group (II) (19 men) did not receive treatment.	rhGH 0.03 mg/kg sc 3 × week for a period of 6 months. Dose of GH was adjusted to maintain IGF-I level in the range of 0.5–1.5U/ml.	Increase in LBM, organ size, muscle size, and skin thickness. Adipose mass (AM) decreased during treatment period. LBM declined and AM increased significantly 3 months after the hGH was stopped.	35% men in group I developed carpal tunnel syndrome, 15% men developed gynecomastia. Adverse effects subsided 3 months after stopping GH injections.
Cohn et al.[33]	83 healthy elderly men with plasma IGF-I level <0.35 U/ml (mean age, 69 years).	Two groups were studied. Group I (n = 62) was treated with hGH and group II (n = 21) was used as control.	Subjects in group I were treated with 0.03 mg/kg body weight hGHsc, 3 × per week for a period of 12 months.	Increase in LBM to 106% of baseline, decrease in adipose mass to 85% of the baseline in group I.	In group 1 31/% of men developed carpal tunnel syndrome and 12% developed gynecomastia. Authors concluded that plasma IGF-I level >1.0 U/ml is associated with adverse effects.

TABLE 4
Clinical Trial of hGH in Elderly Hyposomatomedinemic Men: Outcome Variables as Percentage of Initial Baseline Value

		Month				
	Group	**0**	**6**	**12**	**18**	**21**
Lean body mass	I[a]	100	99.0	104.8[b]	105.7[b]	102.7[b]
	II[c]	100	99.8	99.1	96.0[b]	91.7[b]
Adipose mass	I	100	96.8	86.9[b]	84.8[b]	90.1[b]
	II	100	98.2	102.2	97.8	105.5
Skin thickness (sum of 4 sites)	I	100	98.9	106.4[b]	104.3[b]	
	II	100	99.0	98.0	93.9[b]	
Liver size	I	100	99.2	119[b]	108[b]	
	II	100	98.6	98.3	93.3	
Spleen size	I	100	95.1	116.6[b]	123.0[b]	
	II	100	95.2	101.9	93.2	
Sum of 10 muscle areas	I	100	101.7	111.3[b]	110.6[b]	
	II	100	104.2	96.7	98.3	

[a] Group I, hGH-treated group.
[b] $P < 0.05$ for change from initial baseline value by paired t test.
[c] Group II, no treatment group.
From Rudman, D., et al., *Horm. Res.*, 36 (Supp. 1), 73, 1991. With permission.

after cessation of hormone treatment. Elevation of plasma IGF-I above 1.0 U/ml was associated with increased frequency of carpal tunnel syndrome and gynecomastia. It was concluded that the desirable hormone effects in expanding LBM and reducing AM can be achieved and the undesirable side effects avoided by maintaining the mean IGF-I level in the range of 0.5 to 1.0 U/ml during the treatment period.

C. ADULTS WITH MALNUTRITION, OSTEOPOROSIS, AND OTHER CONDITIONS

Several trials using hGH in elderly with different medical or surgical conditions have been reported.[37–44] Some of the recently published trials for this group of older adults are summarized in Table 6.

In a study by Kaiser et al.[37] nine malnourished men between 64 and 99 years of age were randomly assigned to receive rhGH 100 μg/kg intramuscularly daily for 21 d (treatment group) or a normal saline injection intramuscularly daily for 21 d (control group). The mean baseline IGF-I level was 0.33 U/ml in controls vs. 0.57 U/ml in the treatment group. This study showed an average increase in mid-arm muscle circumference of 0.6 cm, mean weight gain of 2.2 kg, and increased nitrogen retention in the treatment group as compared to the control group. The authors did not report any adverse effects during this short trial.

Suchner et al.[38] studied six malnourished adult men with a mean baseline IGF-I level of 0.91 U/ml, ages 54 to 73 years (mean age, 65.8 years) on parenteral

TABLE 5
Body Composition Responses of Subgroups of Group I Formed According to Mean Intra-treatment IGF-I Level; Group I Carpal Tunnel Cases; and Group II Controls

		At month 12		At month 18	
Subjects (*n*)	Mean IGF-I (U/ml) range during months 7–18	Lean body mass[a]	Adipose mass[a]	Lean body mass[a]	Adipose mass[a]
Group I, no adverse effects *n* = 21)	0.5–1.5	105.9 ± 1.5[b]	85.1 ± 2.9[b]	106.9 ± 1.8[b]	83.3 ± 3.4[a]
Group I, no adverse effects (*n* = 15)	0.5–1.0	107.7 ± 1.6	80.4 ± 3.0	107.0 ± 2.4	81.2 ± 3.9
Group I, no adverse effects (*n* = 6)	1.0–1.5	101.7 ± 2.6[c]	96.8 ± 4.9[c]	102.9 ± 1.8	90.0 ± 5.6
Group I, carpal tunnel syndrome (*n* = 6)	1.0–1.28[d]	103.6 ± 3.2	95.1 ± 6.4		
Group II (*n* = 16)	0.2–0.3	97.7 ± 1.4	107.3 ± 5.0	96.7 ± 0.8	105.1 ± 3.6

Note: Lean body mass and adipose mass are expressed as percentage of initial baseline (average ± SEM).

[a] Expressed as percentage of initial baseline, average ± SEM.
[b] $p < 0.05$ for comparison with group II.
[c] $p < 0.05$ for comparison with (group I, no adverse effects, mean intratreatment IGF-I 0.5–1.0 U/ml).
[d] While 10 group I men developed carpal tunnel syndrome, only six are represented in this table because four dropped out before the body composition measurement at month 12.

From Cohn, L., et al., *Clin. Endocrinol.*, 39, 417, 1993. With permission.

nutrition, and treated them with rhGH. Growth hormone was administered subcutaneously 30μ/d on days 8 to 11 and 60 μ/d on days 12 to 15. The authors demonstrated improvement in nitrogen balance, increase in energy expenditure, and fat oxidation in the treatment group. No untoward side effects were reported.

Four elderly subjects (mean age, 65 yrs) with the history of recent weight loss for a total of 20 d were studied by Binnerts et al.[40] During the entire study period each subject consumed a mixed diet containing 120% of the recommended amount of energy and protein. In addition, each subject was administered rhGH 25 mg/kg/d subcutaneously from days 5 to 8 and 50 mg/kg/d subcutaneously from days 13 to 16. The authors reported significant increase in nitrogen retention of 1.6 and 1.4 g/d, respectively, for each GH treatment period as compared to that in the control periods. Body weight increased 2.3 kg during each treatment period. No adverse effects were reported.

Pape et al.[42] studied the effect of GH in seven malnourished patients with COPD (chronic obstructive pulmonary disease). Subjects with a mean age of 63 years and mean ideal body weight <80% of ideal were studied for a total of 4 weeks. The subjects received a balanced diet of 35 kcal/kg with 1 g of protein per kilogram daily. From day 8 to 28, each subject was administered rhGH 0.05 mg/kg subcutaneously daily. During the GH treatment period subjects showed a mean weight gain of 1.37 kg and a mean increase in nitrogen balance of 2.2 g/d ($p < 0.02$) vs. the diet alone period (week 1). The authors also reported an improvement in maximal inspiratory pressure by 27% after GH treatment. No significant adverse effects were reported. Ward et al.[44] studied 14 elderly subjects undergoing major gastrointestinal surgery. Subjects were randomly assigned to placebo group (mean age, 62 years). Subjects in the treatment group received rhGH 0.1 mg/kg on the day of surgery and the following 6 postoperative days. The authors concluded that rhGH alters post-operative protein and energy metabolism by reducing protein oxidation and increasing fat oxidation. The authors reported elevated blood glucose in the treatment group. No other adverse effects were reported.

A multicenter, randomized, double-blind study in a relatively younger age group (mean age, 42.8 years) investigated whether hGH improves the efficacy of total parenteral nutrition (TPN).[39] Fifteen subjects on TPN were studied (control, nine subjects; treatment group, six subjects) for a period of 14 d. In the treatment group a significant increase in nitrogen, potassium, and phosphate balance occurred as compared to the control group. These authors reported hyperglycemia and peripheral edema in one subject; otherwise no adverse effects were reported. The authors concluded that hGH was well tolerated and significantly enhanced nitrogen retention and improved the efficiency of parenteral nutrient utilization in patients requiring TPN as compared to the control group being fed standard TPN without hGH.

IV. UNANSWERED QUESTIONS ABOUT TREATMENT OF THE ELDERLY WITH GROWTH HORMONE

(1) Who should be classified as being deficient in GH?

Consider the frequency polygons in Figure 1.[15] A common method of defining a normal range for a group is the 95% confidence interval of a healthy reference group. As discussed earlier, 80% of the healthy old men and 90% of the institutionalized old men are hyposomatomedinemic comparing these two groups with the healthy young men.

It is evident from the above discussion that the prevalence of hyposomatomedinemia depends upon the reference ranges used. This issue is important because in clinical endocrinology the benefit:risk ratio of hormone treatment is most favorable in the hormone-deficient subject and is often unfavorable in a person with sufficient endogenous hormone.

TABLE 6
Summary of Clinical Trials Using Human Growth Hormone in Elderly with Medical and Surgical Conditions

Author	Study group	Design	GH regimen	Effects	Adverse effects
Kaiser et al.[37]	Ten malnourished men with body weight >20% or more below average body weight and serum albumin <3.8 g/dl (age, 64–99 years).	Subjects were randomly assigned to treatment group or placebo control group.	Subjects in the treatment group received 100 μg/kg rhGH im, daily for 21 d. The control group received im normal saline injections for 21 d.	Increase in body weight (mean, 2.2 kg), midarm muscle circumference (0.6 cm), and urinary nitrogen retention in the treatment group were reported.	None reported.
Suchner et al.[38]	Six malnourished patients with creatinine height index 56% of predicted; mean serum albumin = 3.4 g/dl and at least 10% weight loss over the last 12 months. All had chronic obstructive pulmonary disease (mean age, 65.8 years).	All patients were receiving parenteral nutrition. It was a 15-d study in which subjects were administered rhGH on day 8–15.	Subjects received rhGH at 30 mg/kg/d sc on days 8–11 and 60 μg/kg/d on days 12–15.	Increase in nitrogen balance, energy expenditure, and fat oxidation were noted.	None reported.
Binnerts et al.[40]	Four subjects (2 men and 2 women) with history of 5 kg weight loss in recent weeks to months with no history of malignancy (mean age, 65 years).	All subjects were on enteral nutritional support and were studied for a period of 20 d.	All subjects received hGH in dosages of 25 and 50 mg/kg/d sc for two 4-d periods.	Increase in nitrogen retention in each treatment period. Body weight also increased probably secondary to water retention.	None reported.

Pape et al.[42]	Seven subjects (4 women and 3 men) with the history of chronic obstructive pulmonary disease and a mean % ideal body weight of 78%, were studied (mean age, 63 years).	It was a 4-week study in which all subjects received hGH for the last 3 weeks.	All subjects received rhGH at a dose of 0.05 mg/kg/d sc for a period of 3 weeks.	Subjects showed an increase in weight gain, nitrogen balance, and maximal inspiratory pressure.	None reported.
Ward et al.[44]	14 subjects who underwent major gastrointestinal surgery were studied for the effect of hGH on post-operative protein and energy metabolism. Mean age for placebo group, 62 years; and for hGH group, 68 years.	Seven patients received placebo and seven patients received hGH for the first 6 post-operative days.	Seven patients received 0.1 mg/kg/d of hGH IM for the first 6 postoperative days.	hGH administration altered post-operative protein and energy metabolism by reducing protein oxidation and increasing fat oxidation with raised rates of whole-body nitrogen turnover.	All patients receiving hGH developed hyperglycemia.

(2) Should the treatment mode be physiologic or pharmacologic?

This question relates to the selection of cases as well as the dose to be used. In the physiologic mode only hormone-deficient individuals would be treated and replacement doses of hormone would be used. The replacement dose is equal to the amount of hormone secreted daily by the reference healthy group. In the pharmacologic mode both hormone-deficient and hormone-sufficient individuals would be treated with doses up to several times the daily secretion of the reference group. The use of thyroxine supplements in a hypothyroid patient is an example of "physiologic mode". Glucocorticords are used both in the "physiologic mode" (as in Addison's disease) as well as in the "pharmacologic mode" (as in systemic lupus erythematosis, psoriasis, asthma, COPD). When a hormone is used in the physiologic mode, beneficial effects are universal and adverse effects are rare, whereas in pharmacologic doses adverse effects are common.

To date, beneficial effects of hGH have been described only in the physiologic mode, i.e., when GH-deficient subjects were treated with replacement doses of hGH.[20–24]

(3) What should be the target range for the IGF-I hormone level in the individuals treated with GH?

In the physiologic mode the objective would be to imitate the natural serum profile of the reference group. Serum IGF-I is the indicator for hGH activity. Insulin-like growth factor-I has no diurnal variation.[45,46] If the reference is healthy young men (HYM), the goal would be to maintain the IGF-I level in the range of 0.5 to 1.5 U/ml (assay of Furlanetto et al.[47]) or 240 to 460 ng/ml (assay of Daughaday et al.[48]). In HYM about two thirds of the daily secretion of GH occurs during the early hours of sleep; therefore, a daily bedtime injection of hGH would be most physiologic.[49,50] This regimen was found to be superior in promoting growth and nitrogen retention in GH-deficient youngsters.[50–53]

(4) What should be the duration of hGH treatment in hyposomatomedinemic elderly subjects?

Both short- and long-term therapeutic goals were proposed. Examples of the short-term therapeutic goals are to accelerate nutritional repletion, to minimize the catabolic effect of trauma or caloric undernutrition, to accelerate wound healing, or to neutralize the tissue-wasting effects of glucocorticords.[37–40,42,43,54,55] Long-term goals include prevention of changes in body composition with aging and the prevention or reversal of loss of bone density.[32,33]

(5) What is the risk of adverse reactions?

Replacement doses in the elderly for 6 to 12 months have caused a substantial incidence of adverse reactions with carpal tunnel syndrome in 24 to 50%, gynecomastia in 9%, and hyperglycemia in 7 to 25%.[33,53] The frequency of these adverse effects was directly related to the intratreatment serum IGF-I concentration, even though this level did not rise above the normal youthful range.[33,53] Using similar replacement doses in children and young to middle-aged adults did not produce any significant adverse effects.[22,23,30,56,57] Therefore, it can be concluded that the elderly are more likely to develop side effects from hGH replacement than children or young to middle-aged adults.

The possibility of delayed undesirable reactions secondary to hGH also needs to be considered. Important possibilities are neoplasia and degenerative arthritis. The neoplastic potential arises for hGH because of the known mitogenic action of IGF-I on most cell types, the presence of IGF-I receptors in some tumors, and the possibly increased frequency of colon and breast tumors in acromegaly.[58–65] The adverse effect of acromegaly on joint cartilage and bone raises the potential that hGH, especially in pharmacologic doses, could aggravate the osteoarthritis present in nearly all old people.[66,67]

(6) Is the declining activity of GH/IGF-I axis in old age adaptive?

Chronic excess of GH in early or mid-adulthood eventually tends to cause diabetes mellitus, osteoarthritis, hypertension, and possibly neoplasm.[68,69] Although the prevalence of each of these conditions progressively rises with advancing age, it can be theorized that the diminishing activity of the GH/IGF-I axis with age reduces the risk of these medical hazards of old age and may be a health trade-off. The cost may be undesirable changes in body composition and bone density, but the gains may be lowered risk of some common geriatric conditions.

(7) Can GHRH or IGF-I substitute for hGH in the treatment of hyposomatomedimenic elderly?

Both GHRH and IGF-I can be synthesized by recombinant DNA technology. The low activity of the GH/IGF-I axis in old age results in part from diminished release of hypothalamic GHRH.[1,2] Furthermore, administration of GHRH for 2 weeks to the elderly raised their IGF-I level to a youthful range.[34,35] Correcting low IGF-I levels with GHRH may have theoretical advantages in comparison to using hGH. The endogenous GH released by GHRH consists of several molecular variants with differing ratios of somatotropic activities, a mixture which may be physiologically superior to exogenous hGH; the latter product corresponds to the major, but not the only component in the naturally secreted mixture.[68] Furthermore, GHRH treatment will leave the inhibitory feedback loop intact whereby any unphysiologic rise in the IGF-I level can shutoff the GH response to GHRH, thereby perhaps protecting against the side effects of hGH common in the elderly.[69] Use of IGF-I may also be advantageous as compared to using hGH in the elderly. While IGF-I reproduces the anabolic, renotropic, and growth-promoting effects of hGH, it does not possess the diabetogenic effect of hGH; instead, IGF-I, like insulin, is hypoglycemic.[20–23,30,32,33,70–73] Kolaczynski and Caro[74] recently reviewed effects of IGF-I therapy in diabetes. This review indicated that patients with type II diabetes who receive IGF-I have improved glucose tolerance and decreased hyperinsulinemia and hypertriglyceridemia. The authors concluded that IGF-I may be a useful adjunct for treatment of diabetes and may even be the drug of choice in some patients with extreme insulin resistance.

V. CONCLUSION

Adverse effects of GH/IGF-I deficiency seem to be a part of the larger picture of the aging endocrine system (e.g., low IGF-II, andropause in males and females, female menopause, low testosterone levels in aging men). Replacement of estrogen in postmenopausal women has proven beneficial. It remains to be seen if replacing hGH or its analogues within elderly men and women, and testosterone in elderly men, will have favorable outcomes with fewer short- and long-term adverse effects.

REFERENCES

1. **Corpas, E., Harman, S. M., and Blackman, M. R.,** Human growth hormone and human aging, *Endocr. Rev.*, 14, 20, 1993.
2. **Cronin, M. J., and Thorner, M. O.,** Basic studies with growth hormone-releasing factor, in *Endocrinology*, Vol. 1, Degroot, L. J., Ed., W.B. Saunders, Philadelphia, 1989, 183.
3. **Reichlin, S., Somatostatin,** *N. Engl. J. Med.*, 309, 1495, 1983.
4. **Hall, R., Schally, A. V., and Evered, D.,** Action of growth hormone release inhibitory hormone in healthy men and acromegaly, *Lancet*, 2, 581, 1973.
5. **Winter, L. M., Shaw, M. A., and Bauman, G.,** Basal plasma growth hormone levels in man: new evidence for rhythmicity of growth hormone secretion, *J. Clin. Endocrinol. Metab.*, 70, 1678, 1990.
6. **Gelato, M. C., Oldfield, E., Loriaux, L., and Merriam, G. R.,** Pulsatile growth hormone secretion in patients with acromegaly and normal men: the effects of growth hormone-releasing hormone infusion, *J. Clin. Endocrinol. Metab.*, 71, 585, 1990.
7. **Underwood, L. E. and VanWyk, J. J.,** Normal and aberrant growth, in *Textbook of Endocrinology*, 8th ed., Wilson, J. D. and Foster, D. W., Eds., W.B. Saunders, Philadelphia, 1992, 1079.
8. **Thorner, M. O., Vance, M. L., Horvath, E., and Kovacs, K.,** The anterior pituitary gland, in *Textbook of Endocrinology*, 8th ed., Wilson, J. D., and Foster, D. W., Eds., W.B. Saunders, Philadelphia, 1992, 221.
9. **Isaksson, O. G., Lindahl, A., Nilsson, A., and Isgaard, J.,** Action of growth hormone: current views, *Acta Paediatr. Scand.*, 343, 1988.
10. **Underwood, L. E., D'Ercole, A. J., Clemmons, D. R., and VanWyk, J. J.,** Paracrine functions of somatomedins, *Clin. Endocrinol. Metab.*, 15, 19, 1986.

11. **Daughaday, W. H. and Rotwein, P.,** Insulin-like growth factors I and II. Peptide, messenger ribonucleic acid and gene structures, serum and tissue concentrations, *Endocr. Rev.,* 10, 168, 1989.
12. **VanWyk, J. J., Underwood, L. E., Hintz, R. L., Clemmons, D. R., Voina, S. J., and Weaver, R. P.,** The somatomedins: a family of insulin-like hormones under growth hormone control, *Recent Prog. Horm. Res.,* 259, 1974.
13. **Clemmons, D. R. and VanWyk, J. J.,** Factors controlling blood concentration of somatomedin-C, *Clin. Endocrinol. Metab.,* 13, 113, 1984.
14. **Rudman, D., Kutner, M. H., Rogers, D. M., Luloin, M. F., Fleming, G. A., and Bain, R. P.,** Impaired growth hormone secretion in the adult population: relation to age and adiposity, *J. Clin. Invest.,* 67, 1361, 1981.
15. **Abbasi, A. A., Drinka, P. J., Mattson, D. E., and Rudman, D.,** Low circulating levels of insulin-like growth factors and testosterone in chronically institutionalized elderly men, *J. Am. Geriatr. Soc.,* 41, 975, 1993.
16. **Keligman, M.,** Age-related alterations of the growth hormone/insulin-like growth factor I axis, *J. Am. Geriatr. Soc.,* 39, 295, 1991.
17. **Florini, J. R., Prinz, P. N., Vitiello, M. V., and Hintz, R. L.,** Somatomedin C levels in healthy young and old men: relationship to peak and 24-hour integrated levels of growth hormone, *J. Gerontol.,* 40, 2, 1985.
18. **Pavlov, E. P., Harman, S. M., Merriam, G. R., Gelato, M. C., and Blackman, M. R.,** Responses of growth hormone (GH) and somatomedin-C to GH-releasing hormone in healthy aging men, *J. Clin. Endocrinol. Metab.,* 62, 595, 1986.
19. **Bennett, A. E., Wahner, H. W., Riggs, B. L., and Hintz, R. L.,** Insulin-like growth factors I and II: aging and bone density in women, *J. Clin. Endocrinol. Metab.,* 59, 701, 1984.
20. **Degerblad, M., Elgindy, N., Hall, K., Sjoberg, H. E., and Thoren, M.,** Potent effect of recombinant growth hormone on bone mineral density and body composition in adults with panhypopituitarism, *Acta Endocrinol., 126, 387, 1992.*
21. **Christianson, J. S., Jørgensen, J. O., Pedersen, S. A., Müller, J., Jørgensen, J., Møller, J., Neickendorf, L., and Skakkebaek, N. E.,** GH-replacement therapy in adults, *Horm. Res.,* 36 (Suppl. 1), 66, 1991.
22. **Degerblad, M., Almkvist, O., Grunditz, R., Hall, K., Kaijser, L., Knutsson, E., Ringertz, H., and Thorén, M.,** Physical and psychological capabilities during substitution therapy with recombinant growth hormone in adults with growth hormone deficiency, *Acta Endocrinol. (Copenhagen),* 123, 185, 1990.
23. **Jørgensen, J. O. L., Pedersen, S. A., Thuesen, L., Jørgensen, J., Ingemann-Hansen, T., Skakkebaek, N. E., Christiansen, J. S.,** Beneficial effects of growth hormone treatment in GH-deficient adults, *Lancet,* 1, 1221, 1989.
24. **Shahi, M., Beshyah, S. A., Hackett, D., Sharp, P. S., Johnston, D. G., and Foale, R. A.,** Myocardial dysfunction in treated adult hypopituitarism: a possible explanation for increased cardiovascular mortality, *Br. Heart J.,* 67, 92, 1992.
25. **Merola, B., Cittadini, A., Colao, A., Longobardi, S., Fazio, S., Sabatini, D., Saccá, L., and Lombardi, G.,** Cardiac structured and function abnormalities in adult patients with growth hormone deficiency, *J. Clin. Endocrinol. Metab.,* 77, 1658, 1993.
26. **Cuneo, R. C., Salomon, F., Willes, C. M., and Sönksen, P. H.,** Skeletal muscle performance in adults with growth hormone deficiency, *Horm. Res.,* 33(Suppl. 4), 55, 1990.
27. **Cuneo, R. C., Salomon, F., Willes, C. M., Hesp, R., and Sönksen, P. H.,** Growth hormone treatment in growth hormone-deficient adults. II. Effects on exercise performance, *J. Appl. Physiol.,* 70(2), 695, 1991.
28. **Whitehead, H. M., Boreham, C., Mellrath, E. M., Sheridan, B., Kennedy, L., Atkinson, A. B., and Hadden, D. R.,** Growth hormone treatment of adults with growth hormone deficiency: results of a 13-month placebo controlled cross-over study, *Clin. Endocrinol.,* 36, 45, 1992.

29. **McGauley, G. A.,** Quality of life assessment before and after growth hormone treatment in adults with growth hormone deficiency, *Acta Paediatr. Scand. (Suppl.)*, 356, 70, 1989.
30. **Salomon, F., Cuneo, R. C., Hesp, R., and Sönksen, P. H.,** The effects of treatment with recombinant human growth hormone on body composition and metabolism in adults with growth hormone deficiency, *N. Engl. J. Med.*, 321, 1797, 1989.
31. **Marcus, R., Butterfield, G., Holloway, L., Gilliland, L., Baylink, D. J., Hintz, R. L., and Sherman, B. M.,** Effects of short term administration of recombinant human growth hormone to elderly people, *J. Clin. Endocrinol. Metab.*, 70, 519, 1990.
32. **Rudman, D., Feller, A. G., Cohn, L., Shetty, K. R. Rudman, I. W., and Draper, M. W.,** Effects of human growth hormone on body composition in elderly men, *Horm. Res.*, 36(Suppl. 1), 73, 1991.
33. **Cohn, L., Feller, A. G., Draper, M. W., Rudman, I. W., and Rudman, D.,** Carpal tunnel syndrome and gynecomastia during treatment of elderly hyposomatomedinemic men with human growth hormone, *Clin. Endocrinol.*, 39, 417, 1993.
34. **Corpas, E., Harman, S. M., Pñeyro, M. A., Roberson, R., and Blackman, M. R.,** Growth hormone (GH)-releasing hormone-(1–29) twice daily reverses the decreased GH and insulin-like growth factor-I levels in old men, *J. Clin. Endocrinol. Metab.*, 75, 530, 1992.
35. **Corpas, E., Harman, S. M., Piñeyro, M. A.** et al., Continuous subcutaneous infusions of growth hormone (GH) releasing hormone 1–44 for 14 days increase GH and insulin-like growth factor-I levels in old men, *J. Clin. Endocrinol. Metab.*, 76, 134, 1993.
36. **Rudman, D.,** Growth hormone, body composition and aging, *J. Am. Geriatr. Soc.*, 33, 800, 1985.
37. **Kaiser, F. E., Silver, A. J., and Morley, J. E.,** The effect of recombinant human growth hormone on malnourished older individuals, *J. Am. Geriatr. Soc.*, 39, 235, 1991.
38. **Suchner, U., Rothkopf, M. M., Stanislaus, G., Elwyn, D. H., Kvetan, V., and Askanzi, J.,** Growth hormone and pulmonary disease: metabolic effects in patients receiving parenteral nutrition, *Arch. Intern. Med.*, 150, 1225, 1990.
39. **Ziegler, T. R., Rombeau, J. L., Young, L. S., Fong, Y., Marano, M., Lawry, S. F., and Wilmore, D. W.,** Recombinant human growth hormone enhances the metabolic efficacy of parenteral nutrition: a double-blind randomized controlled study, *J. Clin. Endocrinol. Metab.*, 74, 865, 1992.
40. **Binnerts, A., Wilson, J. H., and Lamberts, S. W.,** The effects of human growth hormone administration in elderly adults with recent weight loss, *J. Clin. Endocrinol. Metab.*, 67, 1312, 1988.
41. **Clemmons, D. R., Smith-Banks, A., and Underwood, L. E.,** Reversal of diet-induced catabolism by infusion of recombinant insulin-like growth factor-I in humans, *J. Clin. Endocrinol. Metab.*, 75, 234, 1992.
42. **Pape, G. S., Friedman, M., Underwood, L. E., and Clemmons, D. R.,** The effect of growth hormone on weight gain and pulmonary function in patients with chronic obstructive lung disease, *Chest*, 99, 1495, 1991.
43. **Jiang, Z. M., He, G. Z., Zhang, S. Y., Wang, X. R., Yang, N. F., Zhu, Y. Z., and Wilmore, D. W.,** Low dose growth hormone and hypocaloric nutrition attenuate the protein-catabolic response after major operation, *Ann. Surg.*, 210, 513, 1989.
44. **Ward, H. C., Halliday, D., and Sim, J. W.,** Protein and energy metabolism with biosynthetic human growth hormone after gastrointestinal surgery, *Ann.Surg.*, 206, 56, 1987.
45. **Hall, K. and Sara, V. R.,** Somatomedin levels in childhood, adolescence and adult life, *Clin. Endocrinol. Metab.*, 13, 91, 1984.
46. **Clemmons, D. R. and VanWyk, J. J.,** Factors controlling blood concentration of somatomedin C, *Clin. Endocrinol. Metab.*, 13, 113, 1984.
47. **Furlanetto, R. W., Underwood, L. E., VanWyk, J. J., and D'Ercole, A. J.,** Estimation of somatomedin C levels in normals and patients with pituitary disease by radioimmunoassay, *J. Clin. Invest.*, 60, 648, 1977.

48. **Daughaday, W. H., Mariz, I. K., and Blethen, S. L.,** Inhibition of access of bound somatomedin to membrane receptor and immunobinding sites: a comparison of radioreceptor and radioimmunoassay of somatomedin in native and acid-ethanol-extracted serum, *J. Clin. Endocrinol. Metab.*, 51, 781, 1980.
49. **Zadik, Z., Chalew, S. A., McCarter, R. J., Jr., Meistas, M., and Kowarski, A. A.,** The influence of age on the 24-hour integrated concentration of growth hormone in normal individuals, *J. Clin. Endocrinol. Metab.*, 60, 513, 1985.
50. **Ho, K. Y., Evans, W. S., Blizzard, R. M., Veldhuis, J. D., Merrioum, G. R., Samajlik, E., Furlanetto, R., Rogol, A. D., Kaiser, D. L., and Thorner, M. O.,** Effects of sex and age on the 24-hour profile of growth hormone secretion in men: importance of endogenous estradiol concentrations, *J. Clin. Endocrinol. Metab.*, 64, 51, 1987.
51. **Jørgensen, J. O. L., Møller, N., Lauritzen, T., Alberti, K. G. M. M., Ørskov, H., and Christiansen, J. S.,** Evening versus morning injections of growth hormone (GH) in GH-deficient patients: effects on 24-hour patterns of circulating hormones and metabolites, *J. Clin. Endocrinol. Metab.*, 70, 207, 1990.
52. **Rudman, D., Friedes, D., Patterson, J. H., and Gibbas, D. L.,** Diurnal variation in the responsiveness of human subjects to human growth hormone, *J. Clin. Invest.*, 52, 912, 1973.
53. **Jørgensen, J. O.,** Human growth hormone replacement therapy: pharmacological and clinical aspects, *Endocr. Rev.*, 12, 189, 1991.
54. **Herndon, D. N., Barrow, R. E., Kunkel, K. R., Broemeling, L., and Rutan, R. L.,** Effects of recombinant human growth hormone on donor-site healing in severely burned children, *Ann. Surg.*, 212, 424, 1990.
55. **Horber, F. F. and Haymond, M. W.,** Human growth hormone prevents the protein catabolic side effects of prednisone in humans, *J. Clin. Invest.*, 86, 265, 1990.
56. **Frasier, S. O. and Lippe, B. M.,** Clinical review 11. The rational use of growth hormone during childhood, *J. Clin. Endocrinol. Metab.*, 71, 269, 1990.
57. **Aceto, T., Frazier, S. D., Hayles, A. B., Meyer-Bahlburg, H. F. L., Parker, M. L., Munschauer, R., and Di Chiro, G.,** Collaborative study of the effects of human growth hormone in growth hormone deficiency. I. First year of therapy, *J. Clin. Endocrinol. Metab.*, 35, 483, 1972.
58. **Daughaday, W. H.,** The possible autocrine/paracrine and endocrine roles of insulin-like growth factors of human tumors, *Endocrinology*, 127, 1, 1990.
59. **Macauly, V. M., Teale, J. D., Everard, M. J., Joshi, G. P., Smith, I. E., and Millar, J. L.,** Somatomedin-C/insulin-like growth factor-I is a mitogen for human small cell lung cancer, *Br. J. Cancer*, 57, 91, 1988.
60. **Cullen, K. J., Yee, D., Sly, W. S., Perdue, J., Hampton, B., Lippman, M. E., and Rosen, N.,** Insulin-like growth factor receptor expression and function in human breast cancer, *Cancer Res.*, 50, 48, 1990.
61. **Jammes, H., Payrat, J. P., Ban, E., Vilain, M. O., Haour, F., Djiane, J., and Bonneterre, J.,** Insulin-like growth factor I receptors in human breast tumor: localization and quantification by histo-autoradiographic analysis, *Br. J. Cancer*, 66, 248, 1992.
62. **Osborne, C. K., Clemmons, D. R., and Arteaga, C. L.,** Regulation of breast cancer growth by insulin-like growth factors, *J. Steroid Biochem. Mol. Biol.*, 37, 805, 1990.
63. **Ezzat, S., and Melmed, S.,** Clinical review 18: are patients with acromegaly at increased risk for neoplasia?, *J. Clin. Endocrinol. Metab.*, 72, 245, 1991.
64. **Brunner, J. E., Johnson, C. C., Zafar, S., Peterson, E. L., Brunner, J. F., and Mellinger, R. C.,** Colon cancer and polyps in acromegaly: increased risk associated with family history of colon cancer, *Clin. Endocrinol.*, 32, 65, 1990.
65. **Bengtsson, B.-Å., Edén, S., Ernest, I., Odén, A., and Sjögren, B.,** Epidemiology and long-term survival in acromegaly. A study of 166 cases diagnosed between 1955 and 1984, *Acta Med. Scand.*, 223, 327, 1988.
66. **Bluestone, R., Bywaters, E. G., Hartog, M., Holt, P. J. L., and Hyde, S.,** Acromegalic arthropathy, *Ann. Rheum. Dis.*, 30, 243, 1971.

67. **Detenbeck, L. C., Tressler, H. A., O'Duffy, J. D., and Randall, R. V.,** Peripheral joint manifestations of acromegaly, *Clin. Orthoped. Rel. Res.,* 91, 119, 1973.
68. **Jadresic, A., Banks, L. M., Child, D. F., Diamant, L., Doyle, F. H., Fraser, T. R., and Joplin, G. F.,** The acromegaly syndrome. Relation between clinical features, growth hormone values and radiological characteristics of the pituitary tumors, *Q. J. Med.,* 51, 189, 1982.
69. **Nabarro, J. D. N.,** Acromegaly, *Clin. Endocrinol. (Oxford),* 26, 481, 1987.
70. **Guler, H. P., Zapf, J., and Froesch, E. R.,** Short-term metabolic effects of recombinant human insulin-like growth factor-I in healthy adults, *N. Engl. J. Med.,* 317, 137, 1987.
71. **Jacob, R., Barrett, E., Plewe, G., Fagin, K. D., and Sherwin, R. S.,** Acute effects of insulin-like growth factor I on glucose and amino acid metabolism in the awake fasted rat. Comparison with insulin, *J. Clin. Invest.,* 83, 1717, 1989.
72. **Guler, H. P., Schmid, C., Zapf, J., and Froesche, E. R.,** Effects of recombinant insulin-like growth factor-I on insulin secretion and renal function in normal human subjects, *Proc. Natl. Acad. Sci. U.S.A.,* 86, 2868, 1989.
73. **Boulware, S. D., Tamborlane, W. V., Sherwin, R. S., and Matthews, L. S.,** Diverse effects of insulin-like growth factor-I on glucose, lipid and amino acid metabolism, *Am. J. Physiol.,* 262, E130, 1992.
74. **Kolaczynski, J. W. and Caro, J. F.,** Insulin like growth factor 1 therapy in diabetes: physiologic basis, clinical benefits and risks, *Ann. Intern. Med.,* 120, 47, 1994.

Chapter 10

GROWTH HORMONE EFFECTS ON THE IMMUNE SYSTEM

Douglas A. Weigent and J. Edwin Blalock

TABLE OF CONTENTS

I. INTRODUCTION

The importance of growth hormone (GH) in the stimulation of growth has been appreciated for many years. There is also ample evidence to suggest a role for GH as a modulator of the immune system.[1–3] This review, summarizes the current aspects of immunoregulation by GH and points out areas requiring further investigation.

II. SYNTHESIS AND SECRETION OF GROWTH HORMONE BY THE PITUITARY

Recent studies have increased our understanding of the complexity of the structure, function, and production of GH.[4] Growth hormone is a single-chain polypeptide of 191 amino acids which is produced by the somatotropic cells of the anterior pituitary. The GH gene is approximately 2.5 kb in length and consists of five exons separated by four introns. Human chromosome 17 contains two GH genes termed hGH-N and hGH-V; the corresponding proteins are highly homologous and have 13 differences out of 191 amino acids dispersed throughout the polypeptide chain. Both genes give rise to variants in which the hGH-N gene is expressed in the pituitary gland while the hGH-V gene is primarily expressed in the placenta. Transcripts of the hGH-N gene are spliced into two different messenger RNAs (mRNAs) that code for a 22- and a 20-kDa GH containing a deletion of a 15 amino acid residue encompassing residues 32 to 46. A number of

0-8493-8384-6/95/$0.00+$.50

post-translational modifications are known to occur, including deamidation, acylation, and glycosylation. Overall, the number of GH forms identified in plasma exceeds 100.[4] Future studies should aim at developing antibodies that are specific for each variant with the goal of identifying their various biological roles.

The cloning of the GH gene has enabled a search for regulatory factor binding sites on the promoter region. The results show that the promoter binds several ubiquitous transcription factors, including NF-1, AP-2, USF, and SP1, and a specific factor, GHF-1 or pit-1.[5] Growth hormone releasing hormone (GHRH) is thought to stimulate GH gene expression via a cyclic adenosine monophosphate-dependent induction of pit-1. Dexamethasone and T_3 also stimulate rat GH gene transcription, most likely by interacting at specific sites within the promoter region. The liver is the predominant source of insulin-like growth factor-I (IGF-I), which inhibits GH secretion in the pituitary and GHRH secretion in the hypothalamus.

Pituitary GH secretion is primarily regulated by the interaction of two hypothalamic peptides, GHRH and somatostatin.[6] In the rat GHRH occurs as a 43 amino acid peptide which binds to specific cell surface receptors, and via cAMP-dependent pathways stimulates GH synthesis and release. The main action of hypothalamic somatostatin is to inhibit the synthesis and release of GH. The daily integrated concentrations of GH are greater in women than men and GH pulses occur with an average frequency of about 13/d in both sexes. The available evidence suggests that spontaneous GH release is under somatostatin inhibitory control and that secretory pulses occur during periodic decreases in somatostatin secretion. Human studies have shown that GH pulse amplitudes are increased during slow wave sleep and after exercise. In addition, depression, hyperglycemia, and obesity can reduce basal and stimulated secretion of GH. Data from studies in rats and humans suggest that age-related reductions in GH secretion result from an increase in somatostatin tone and a decrease in GHRH secretion.[7]

III. LYMPHOCYTES AS A SOURCE OF GROWTH HORMONE

Numerous studies indicate that in addition to the pituitary serving as a site of GH production, normal lymphocytes as well as lymphoid cell lines are capable of synthesizing and releasing GH.[8] Our studies in rats and those by others in humans indicate that the GH molecule produced by cells of the immune system is structurally identical to pituitary GH. In particular, the GH molecule from lymphocytes was shown to be similar, if not identical, to pituitary GH in terms of molecular weight, immunogenicity, and bioactivity.[9] Restriction enzyme analysis of complementary DNA (cDNA) synthesized from purified lymphocyte mRNA is similar to the cDNA obtained from pituitary RNA.[10] In more recent studies we cloned and sequenced a cDNA from lymphocytes corresponding in sequence to exons C, D, and E of the pituitary. Overall, the data support the idea that the lymphocyte GH protein is identical to the 22-kDa protein secreted by the pituitary.

TABLE 1
Growth Hormone Axis

Organ	Hormone	Action
Hypothalamus	GHRH	Stimulates GH transcription and secretion
	Somatostatin	Inhibits GH secrection
Pituitary	GH	Stimulates IGF-I production Antagonizes insulin action
Liver and extrahepatic tissues	IGF-I	Stimulates cell replication and bone growth
Immune system	GH	Stimulates cell replication Stimulates lymphocyte IGF secretion
	GHRH	Stimulates GH synthesis and proliferation in lymphocytes
	IGF	Stimulates cell proliferation Inhibits lymphocyte GH symthesis
	Somatostatin	Inhibits lymphocyte GH synthesis

The synthesis and secretion of GH by immune cells is similar to that described for the pituitary (Table 1). Thus, GHRH stimulates GH synthesis whereas IGF-I and somatostatin seem to inhibit lymphocyte GH synthesis. An important difference with the pituitary appears to be that the gene in lymphocytes is under tonic suppression.[9] The removal of spleen or thymic tissues from animals results in rapid, spontaneous gene transcription; maximum levels of GH mRNA are reached after only 8 h without any obvious stimulation. In another report exogenous GH was suggested to augment endogenous GH secretion from nonstimulated and phytohemagglutinin-stimulated peripheral blood mononuclear cells.[11] The expression of pit-1 was documented in human and rat lymphoid tissues and cell lines.[12] Our data from studies of dwarf mice suggest near-normal expression of GH in lymphocytes and raise the possibility that another protein may substitute for pit-1 in cells of the immune system.

Cells of the immune system also appear to have the ability to produce hypothalamic releasing factors similar in structure and function to those originally identified in the neuroendocrine system.[13] Numerous studies were conducted by us and others to determine whether molecules involved in pituitary GH regulation could be produced by cells of the immune system and play a role in the regulation of lymphocyte GH. In this regard we demonstrated that cells of the immune system also produce GHRH.[13] The nucleic acid studies in the rat showed that lymphocyte GHRH-related RNA was polyadenylated and of the same molecular mass as hypothalamic GHRH RNA.[10] The protein data obtained using antibody affinity chromatography, followed by size separation on high performance liquid chromatography columns demonstrated a 5-kDa molecular weight species that was *de novo* synthesized. The purified lymphocyte GHRH molecule could bind to the GHRH receptor and stimulate an increase in GH mRNA synthesis in lymphocytes and pituitary cells.[13] We also showed that cells of the immune system have receptors for GHRH similar to those described on pituitary cells.[14] The studies on the binding of GHRH to spleen and thymus cells showed that it was saturable, of high affinity, and specific. The treatment of leukocytes with GHRH rapidly increased the intracellular levels of calcium, thymidine incorporation, and the levels of GH RNA in the cytoplasm. In addition to GHRH, the hypothalamic-inhibiting hormone somatostatin was identified in platelets, mononuclear leukocytes, mast cells, and polymorphonuclear leukocytes.[15] Although the amino acid sequences have not been completed, the immune-derived somatostatin appears to be larger than the authentic neuropeptide and to have a similar but not identical amino acid composition.

Finally, considerable data has been published documenting the ability of cells of the immune system to produce IGF-I-like molecules. In particular, activated alveolar macrophages release an IGF-I molecule.[16] Human B lymphocytes, transformed with Epstein-Barr virus, produce IGF-I *in vitro* and respond to GH by significantly increasing IGF-I secretion.[17] Complementary DNA probes for rat IGF-I in Northern blot hybridizations demonstrated that spleen and thymus tissues contain mRNA for IGF-I.[18] The assay of basal- and GH-stimulated conditioned media demonstrated low levels of IGF-I from normal T cells. Our own studies identified leukocytes positive for IGF-I production by direct immunofluorescence.[19] Results using immunoaffinity purification, high performance liquid chromatography, and a fibroblast proliferation bioassay suggest that the *de novo* synthesized leukocyte-derived IGF-I is similar in molecular weight, antigenicity, and bioactivity to serum IGF-I. The levels of leukocyte-derived IGF-I increase after treatment of lymphoid cells with GH.[19] In conclusion, cells of the immune system can serve as a source of neuroendocrine hormones and neuropeptides. The molecules are the same or very similar, and they are regulated similarly to their neuroendocrine counterparts.

The idea that neuroendocrine hormones produced by lymphocytes function in mostly paracrine or autocrine rather than endocrine roles is supported by the low levels of hormone produced by immune cells. In the case of GH approximately 10% of lymphoid cells are positive by immunofluorescence, whereas only 0.1% of cells are secreting GH as determined by the reverse hemolytic plaque assay.[20] Thus, much more of the lymphocyte hormone appears to remain intracellular and may function in an intracellular fashion similar to prolactin or in an autocrine manner.[19,21,22] In other studies we analyzed the production of GH mRNA and the secretion of GH by purified subpopulations of rat lymphoid cells.[23] The results show that mononuclear leukocytes from various tissues, including spleen, thymus, bone marrow, Peyer's patches, and peripheral blood cells, all have the ability to produce GH mRNA and secrete GH. Data obtained from cells separated by adherence, nylon wool columns, and positive and negative sorting with monoclonal antibodies that define B, monocyte, T helper, and T cytotoxic cells show that several different cell types have the ability to produce GH mRNA. These results suggest that B cells and macrophages produce more GH mRNA than do T cells.

Natural killer cells also produce detectable amounts of GH mRNA. Most interesting was the finding of a remarkably low level of GH mRNA expression in T suppressor cells relative to T helper cells. This supports the idea that GH generally promotes or has a positive effect on immune cell functions. In addition to these findings, we also showed that the same cells that produce GH also produce IGF-I.[24] Taken together the results suggest that an autocrine regulatory circuit may be important for the production of leukocyte-derived GH and IGF-I within the immune system. The relevance of these observations to humoral and cellular immunity remains to be determined.

IV. BINDING PROTEINS FOR GROWTH HORMONE

After secretion, GH associates with circulating GH binding proteins (GHBP).[25] The recent discovery of circulating GHBP in human and rabbit plasma has offered new perspectives on GH action. The GHBP was discovered in the process of characterizing circulating GH forms and GH receptor-like proteins in serum. Growth hormone circulates in the blood bound to two different binding proteins. One is a high-affinity GHBP which complexes almost 90% of the 22 kDa GH protein and a second low-affinity GHBP which complexes the minor 20 kDa GH variant. In subjects with Laron syndrome who lack GH receptors from a gene mutation serum binding protein levels vary from 0 to 2% of normal values. There is a single gene encoding the GH receptor and no gene for the GHBP has been identified. In humans the GHBP is thought to arise from receptor cleavage, while in rodents two mRNA species were identified consistent with alternative splicing. Because the liver is the organ with the highest GH receptor concentrations, it may be the main source of GHBP. However, GH receptors are demonstrable on a variety of tissues, including lymphocytes, and very little is known about the mechanism and secretion of GHBP. Plasma levels of GHBP fluctuate little throughout the day and in adults, neither GH deficiency or hypersecretion results in substantially altered GHBP levels. The high-affinity GHBP inhibits binding of GH to receptors *in vitro* while GHBP enhances the growth-promoting actions of GH *in vivo*.[25] This paradox can be explained by the ability of the GHBP to prolong the half-life of GH. Although the physiologic roles of GHBP remain to be defined, it seems very likely that GHBP will be found to have local effects, for example, at sites of inflammation, that influence the binding and action of GH.

V. LYMPHOCYTES AND GROWTH HORMONE RECEPTORS

The idea that lymphocytes have receptors for GH is supported by biochemical, molecular, and biological data.[26] The receptors for GH belong to a family of hemopoietic growth factor receptors. Several members of this family include prolactin, interleukin-2(L-2), IL-3, IL-4, IL-6, and erythropoietin. In the case of GH the receptor consists of a single polypeptide chain of 620 residues with a single transmembrane domain. The receptors in this family are rich in cysteines in their extracellular domains, a theme that may be important in protein to protein and cell to cell interactions. However, the intracellular domains of these receptors show no homology with each other or other receptors which have domains corresponding to known signaling mechanisms. It appears, however, that a number of growth factor receptors, including GH, act through activation of a tyrosine kinase.[27] Ilondo et al.[28] recently showed in lymphocytes that GH stimulates the rapid tyrosine phosphorylation of two proteins with a molecular weight of 120 and 93 kDa in a time- and dose-dependent manner in a human lymphocyte line, IM9. By treating IM9 cells with mutant GH proteins it was shown that the phosphorylation of the 120 and 93 kDa proteins is maximal when two molecules of receptor bind to one molecule of GH. The same receptor surface appears to bind to two different binding sites on the opposite sides of the four-helical bundle of the hormone. Overall, the structure supports the notion that dimerization of the extracellular domain would lead to dimerization of the intracellular domain and consequent signaling.[29] After the binding of GH to its receptor, the GH receptor complex is internalized and activation of intracellular signals, including tyrosine kinase, occurs. At the same time, the cell membrane becomes refractory and GH binding decreases markedly until a new supply of receptors arrives from *de novo* synthesis or from recycling of processed receptors. The data suggest exocytosis of intact hormone via recycling endosomes as well as degradation in the lysosomes. In humans the GHBP is thought to derive from the membrane receptor by proteolytic cleavage. Such a process could take place in the endosomes and provide a mechanism for release of the high-affinity GHBP. Overall, very little receptor recycling occurs and a rapid receptor synthetic process is required to maintain the cell surface complement of GH receptors.[30] Future studies on the intracellular fate of lymphocyte GH and its receptor will show whether these pathways play important roles in initiating, maintaining, or catalyzing biological responses to GH.

VI. *IN VITRO* EFFECTS OF GROWTH HORMONE ON THE IMMUNE RESPONSE (Table 2)

The expression of GH and its receptor on lymphoid cells would indicate that GH exerts physiological effects on the immune system.[2,4] The potential role of GH in immunoregulation was demonstrated for numerous immune functions and on isolated cells *in vitro* with concentrations as low as 0.1 ng/ml. DNA and RNA synthesis in the spleen and thymus of normal and hypophysectomized rats increased after GH treatment. Growth hormone also affects hematopoiesis by stimulating neurophil differentiation and augmenting erythropoiesis. Treatment via GH increases proliferation of bone marrow cells and influences thymic development. Proliferative responses of lymphoid cells are greater when treated with GH *in vitro.* Growth hormone affects the functional activity of cytolytic cells, including T lymphocytes and natural killer cells,[31] and was necessary for T lymphocytes to

TABLE 2
Effects of GH on Immune Cell Functions

Immune function	Effect
Thymic size and DNA synthesis	Increases
Erythropoiesis	Augments
Neutrophil differentiation	Augments
Antibody synthesis	Augments
Skin graft survival	Decreases
Basal lymphocyte proliferation	Augments
Lectin-induced T cell proliferation	Stimulates
IL-1 and IL-2 synthesis	Increases
Cytotoxic T cell activity	Increases
Natural killer cells	Augments
Tumor necrosis factor-α and superoxide anion release by macrophages	Stimulates
Thymulin production	Stimulates

develop cytolytic activity against an allogeneic stimulus in serum-free media.[32] The cytolytic activity of natural killer cells is reduced by hypophysectomy, and this effect can be partially reversed by administration of GH *in vivo.* Growth hormone was shown to stimulate the production of superoxide anion formation from macrophages.[33]

Several recent studies raise the possibility that GH may be exerting its effects through another hormone receptor or the induction of another growth factor. In addition to data showing the displacement of GH from lymphocytes by prolactin, it has now been clearly shown that the binding of GH to neutrophils to stimulate superoxide anion secretion occurs through the prolactin receptor.[34] When bone marrow cells are cultured in medium containing granulocyte/macrophage colony stimulating factor and GH, a significant enhancement occurs in myeloid colony formation. The addition of antiserum to the IGF-I receptor was seen to block the response to GH. Also, the GH-stimulated growth of virus-transformed T lymphoblast cell lines could be blocked by antibodies to IGF-I.[35] Infusions of IGF-I reverse thymic atrophy in hypophysectomized rats and increase thymocyte proliferation.[36] It is not clear whether GH directly influences intra- or extrathymic development or acts indirectly by augmenting the synthesis of thymulin or IGF-I. More recent work has shown that stimulation of T cells with IL-2 induced expression of IGF-I receptors is consistent with the idea that IGF-I is important in the activation of T cells.[37] Tissue IGF-I levels rise in response to GH in hypophysectomized rats and IGF-I levels in humans were found to increase in various extrahepatic tissues, including T lymphocytes. These observations suggest that GH stimulates local production of IGF-I, which acts to promote tissue growth and action in a paracrine/autocrine fashion. Many of the leukocyte functions stimulated by GH can also be augmented by IGF-I treatment. Thus, it appears that the action of GH on cells of the immune system may be mediated in part by IGF-I.

Growth hormone releasing hormone was shown to modulate the immune system, increasing GH mRNA in leukocytes and to cause a twofold increase in thymidine and uridine incorporation of nonstimulated lymphocytes.[14] It was also shown to stimulate lymphocyte proliferation, to inhibit natural killer activity, and to inhibit the chemotactic response.[38] Collectively, these results indicating the ability of lymphocytes to express GH, GHRH, and IGF-I receptors and their ability to respond to GH, GHRH, and IGF-I illustrates the potential of these molecules to function as growth factors as well as immunoregulatory molecules capable of influencing humoral and cellular aspects of the immune system.

Most studies have examined the effect of exogenously added GH on selected immune responses. The presence of lymphocyte-derived GH produced by cells of the immune system makes it more difficult to study. Nevertheless, we have employed antisense oligodeoxynucleotides (ODNs) and antibodies to block the endogenous activity of GH.[22] These studies showed that treatment of rat lymphocytes with a specific GH antisense ODN decreased the amount of leukocyte GH synthesized and lymphocyte proliferation. The antisense ODN growth inhibition could be prevented by complementary GH sense ODN and reversed by the exogenous addition of rat GH. In studies with antibodies to GH we were able to measure a twofold decrease in the number of cells positive for IGF-I. This strongly supports an important role for endogenously produced GH in the induction of leukocyte-derived IGF-I.[19] Taken together our findings support the idea that GH axis hormones function in a paracrine/autocrine loop. More studies are needed to identify the induction mechanisms to delineate further the roles of these hormones in immunity.

VII. *IN VIVO* EFFECTS OF GROWTH HORMONE ON THE IMMUNE RESPONSE (Table 2)

The idea that GH influences cells of the immune system *in vivo* was initially shown when mice injected with antibody to GH developed thymic atrophy. In addition, dwarf mice showed a reduced ability to synthesize antibodies. Although a number of immune events are abnormal in GH-deficient humans, they are generally not considered to be immunodeficient. The changes in humans include thymic hypoplasia, reduced activity of natural killer cells, and reduced ability of lymphoid cells to respond or stimulate cells in an allogeneic mixed lymphocyte reaction.[2,39] In contrast to humans, many studies have been done in both normal and GH-deficient animals which demonstrate the ability of GH to modulate immune function.[2–4] Overall, injections of GH increase thymic size, stimulate thymocyte proliferation, and induce expression of the DNA-binding protein c-*myc* in hypophysectomized rats. Growth hormone augments antibody synthesis and skin-graft rejection when injected into hypopituitary animals. It can reverse the leukopenia caused by stress. Injected *in vivo,* GH increases both basal and lectin-induced proliferative responses from spleen cells of aged rats.[40] Other studies *in vivo* have shown that GH can stimulate the production of IL-1, IL-2, tumor necrosis factor-α, and thymulin; induce the cytotoxic activity of natural killer cells, and restore the normal architecture of the thymus in aged animals.[2,3] More recently it was shown that GH promotes lymphocyte engraftment in immunodeficient mice.[41] Overall it appears that GH has a pleiotropic effect upon the thymus, functionally altering the lymphoid compartment as well as thymulin production by thymic epithelial cells. Animal studies suggest that infusion of IGF-I may stimulate growth and restore the weight of atrophied organs of hypophysectomized animals.[2–4] The only recent availability of recombinant human IGF-I has limited the number of studies in humans. Alternative methods of treatment with GHRH and superactive analogues are currently being investigated to improve its use as a therapeutic agent.

The role of lymphocyte-derived GH *in vivo* has not yet been investigated well. Most of the data point toward a paracrine/autocrine role within the immune system, but an endocrine role cannot be ruled out. The possibility that neuroendocrine hormones secreted by lymphocytes may function in an endocrine fashion and influence the function of other cells has major implications for the role of leukocyte-derived peptides. This was tested once in the pituitary adrenal system with adrenocorticotropin and the results were controversial. We conducted studies to examine whether syngeneic transfer of GH-producing leukocytes could stimulate the growth of dwarf mice. The results showed that normal spleen cells alone or spleen cells treated with GHRH did not appear to significantly stimulate the growth of dwarf mice, although a trend toward growth was noted.[42] Spleen cells activated *in vitro* with concanavalin A or lipopolysaccharide and then transferred to dwarf mice or thymus cells alone were also without effect, whereas GH alone stimulated growth as expected. Serum levels of IGF-I and IGF-I liver RNA were undetectable in control dwarf mice and dwarf mice receiving spleen cells, whereas serum levels of IGF-I increased after treatment of dwarf mice with GH. The immune system of dwarf mice receiving spleen cells, however, was significantly altered. Spleen cells from dwarf animals showed enhanced immunoglobulin production, IL-6, IL-2, and IGF-γ production, whereas no significant change was apparent in natural killer cell activity. Further proof that lymphocyte-derived GH may be responsible for the activation of the immune system in the dwarf mouse should be obtained by evaluating immune parameters in dwarf mice treated with antibodies to GH to neutralize the GH produced by lymphocytes.

The availability of recombinant hGH has resulted in renewed interest in GH therapy. Growth hormone treatment improves physical strength and psychological well being.[43] It is known that the immune system can be conditioned and that stressed animals have altered susceptibility to tumors.[44] The nervous and immune systems interact and neurochemical and physiological changes in brain cells as a result of GH treatment may result in the activation of the immune system. Growth hormone also stimulates T cell development, which suggests that it may be useful for the treatment of T cell deficiencies.[41] A source of concern has been the association of leukemia with GH treatment. Despite the observed mitogenic activity of GH and IGF-I and the ability of GH to induce c-*myc* protooncogene *in vivo,* the present data, at least in the U.S., indicates that GH is not an added leukemogenic risk to patients with GH deficiency who do not have coexisting risk factors.[41,44] Growth hormone was also used to stimulate growth in children with renal allografts where the catabolic effects of steroids depress growth as well as the immune system.[45] It is not exactly clear that the GH may stimulate the immune system and induce an acute or chronic rejection crisis. Growth hormone was also shown to exert hematopoietic growth-promoting effects *in vivo* and partially counteract the myelosuppressive effects of azidothymidine.[46] These results suggest the need for a controlled study to evaluate the efficacy of GH in the management of the wasting disease in AIDS patients.

The evidence to date shows that GH is an important regulator of the immune system. This is supported by a number of *in vitro* and *in vivo* studies reporting the effects of GH on immunity. Also, it was found that GH, GHRH, and IGF-I and their receptors are produced in the immune system. A great deal more research will be necessary to dissect the role(s) of GH in lymphocytes as well as its potential role in health, disease, and aging.

ACKNOWLEDGMENT

The research was supported in large part by grants from the National Institute of Neurology and Communicative Disorders (RO1 NS24636) and the National Institute of Arthritis, Diabetes, Digestion and Kidney Disease (RO1 DK38024). The authors are indebted to former graduate students, postdoctoral fellows, and faculty colleagues for their contributions to the work reviewed here. We also thank Diane Weigent for excellent editorial assistance and typing the manuscript.

REFERENCES

1. **Gala, R. R.,** Prolactin and growth hormone in the regulation of the immune system, *Soc. Exp. Biol. Med.,* 37, 513, 1991.
2. **Kelley, K. W.,** Growth hormone, lymphocytes, and macrophages, *Biochem. Pharmacol.,* 35, 705, 1989.
3. **Gelato, M. C.,** Growth hormone-insulin-like growth factor I and immune function, *Trends Endocrinol. Metab.,* 4, 106, 1993.
4. **Baumann, G.,** Growth hormone heterogeneity: genes, isohormones, variants, and binding proteins, *Endocr. Rev.,* 12, 424, 1991.
5. **Rousseau, G. G.,** Growth hormone gene regulation by transacting factors, *Horm. Res.,* 37, 88, 1992.
6. **Cronin, M. J. and Thorner, M. O.,** Basic studies with growth hormone-releasing factor, in *Endocrinology,* Vol. 1, DeGroot, L. J., Ed., W. B. Saunders, Philadelphia, 1989, 183.
7. **Corpas, E., Harman, M., and Blackman, M.,** Human growth hormone and human aging, *Endocr. Rev.,* 14, 20, 1993.
8. **Blalock, J. E.,** A molecular basis for bidirectional communication between the immune and neuroendocrine systems, *Physiol. Rev.,* 69, 79, 1989.

9. **Weigent, D. A., Baxter, J. B., Wear, W. E., Smith, L. R., Bost, K. L., and Blalock, J. E.,** Production of immunoreactive growth hormone by mononuclear leukocytes, *FASEB J.,* 2, 2812, 1988.
10. **Weigent, D. A., Riley, J. E., Galin, F. S., LeBoeuf, R. D., and Blalock, J. E.,** Detection of GH and GHRH-related messenger RNA in rat lymphocytes by the polymerase chain reaction, *Proc. Soc. Exp. Biol. Med.,* 198, 643, 1991.
11. **Hattori, N., Shimatsu, A., Sugita, M., Kumagai, S., and Imura, H.,** Immunoreactive growth hormone (GH) secretion by human lymphocytes: Augmented release by exogenous GH, *Biochem. Biophys. Res. Commun.,* 168, 396, 1990.
12. **Delhase, M., Vergani, P., Malur, A., Hooghe-Peters, E. L., and Hooghe, R. J.,** The transcription factor Pit-1/GHF-1 is expressed in hemopoietic and lymphoid tissues, *Eur. J. Immunol.,* 23, 951, 1993.
13. **Weigent, D. A. and Blalock, J. E.,** Immunoreactive growth hormone releasing hormone in rat leukocytes, *J. Neuroimmunol.,* 29, 1, 1990.
14. **Guarcello, V., Weigent, D. A., and Blalock, J. E.,** Growth hormone releasing factor receptors on lymphocytes, *Cell. Immunol.,* 136, 291, 1991.
15. **Aquila, M. C., Dees, W. L., Haensly, W. E., and McCann, S. M.,** Evidence that somatostatin is localized and synthesized in lymphoid organs, *Proc. Natl. Acad. Sci. U.S.A.* 88, 11485, 1991.
16. **Rom, W. N., Basset, P., Fells, G. A., Nukiwa, T., Trapnell, B. C., and Crystal, R. G.,** Alveolar macrophages release an insulin-like growth factor I-type molecule, *J. Clin. Invest.,* 82, 1685, 1988.
17. **Merimee, T. J., Grant, M. B., Broder, C. M., and Cavalli-Sforza, L. L.,** Insulin-like growth factor secretion by human B-lymphocytes: a comparison of cells from normal and pygmy subjects, *J. Clin. Endocrinol. Metab.,* 69, 978, 1989.
18. **Murphy, L. J., Bell, G. I., and Friesen, H. G.,** Tissue distribution of insulin-like growth factor I and II messenger ribonucleic acid in the adult rat, *Endocrinology,* 120, 1279, 1987.
19. **Baxter, J. B., Blalock, J. E., and Weigent, D. A.,** Characterization of immunoreactive insulin-like growth factor-I from leukocytes and its regulation by growth hormone, *Endocrinology,* 129, 1727, 1991.
20. **Kao, T.-L., Supowit, S. C., Thompson, E. A., and Meyer, W. J., III,** Immunoreactive growth hormone production by human lymphocyte cell lines, *Cell. Mol. Neurobiol.,* 12, 483, 1992.
21. **Clevenger, C. V., Sillman, A. L., and Prystowsky, M. B.,** Interleukin-2 driven nuclear translocation of prolactin in cloned T-lymphocytes, *Endocrinology,* 127, 3151, 1990.
22. **Weigent, D. A., LeBoeuf, R. D., and Blalock, J. E.,** An antisense oligonucleotide to growth hormone mRNA inhibits lymphocyte proliferation, *Endocrinology,* 128, 2053, 1991.
23. **Weigent, D. A. and Blalock, J. E.,** The production of growth hormone by subpopulations of rat mononuclear leukocytes, *Cell. Immunol.,* 135, 55, 1991.
24. **Weigent, D. A., Baxter, J. B., and Blalock, J. E.,** The production of growth hormone and insulin-like growth factor-I by the same subpopulation of rat mononuclear leukocytes, *Brain, Beh. Immunol.,* 6, 365, 1992.
25. **Baumann, G.,** Growth hormone-binding proteins, *Proc. Soc. Exp. Biol. Med.,* 202, 392, 1993.
26. **Kiess, W. and Butenandt, O.,** Specific growth hormone receptors on human peripheral mononuclear cells: reexpression, identification and characterization, *J. Clin. Endocrinol. Metab.,* 60, 740, 1985.
27. **Kelly, P. A., Dijiane, J., Postel-Vinay, M.-C., and Edery, M.,** The prolactin/growth hormone receptor family, *Endocr. Rev.,* 12, 235, 1991.
28. **Ilondo, M. M.,** Vanderschueren-Lodeweyckx, M., Courtoy, P. J., and De Meyts, P., Cellular processing of growth hormone in IM-9 cells: evidence for exocytosis of internalized hormone, *Endocrinology,* 130, 2037, 1992.
29. **DeVos, A. M., Ultsch, M., and Kossiakoff, A. A.,** Human growth hormone and extracellular domain of its receptor: crystal structure of the complex, *Science,* 255, 306, 1992.
30. **Roupas, P. and Herington, A. C.,** Cellular mechanisms in the processing of growth hormone and its receptor, *Mol. Cell. Endocrinol.,* 61, 1, 1989.

31. **Saxena, Q. B., Saxena, R. K., and Adler, W. H.,** Regulation of natural killer activity in vivo. III. Effect of hypophysectomy and growth hormone treatment on the natural killer activity of the mouse spleen cell population, *Int. Arch. Allergy Appl. Immunol.,* 67, 169, 1982.
32. **Snow, E. C., Feldbush, T. L., and Oaks, J. A.,** The effect of growth hormone and insulin upon MLC responses and the generation of cytotoxic lymphocytes, *J. Immunol.,* 126, 161, 1981.
33. **Edwards, C. K., III, Ghiasuddin, S. M., Schepper, J. M., Yunger, L. M., and Kelley, K. W.,** A newly defined property of somatotropin: priming of macrophages for production of superoxide anion, *Science,* 239, 769, 1988.
34. **Fu, Y.-K., Arkins, S., Fuh, G., Cunningham, B. C., Wells, J. A., Fong, S., Cronin, M. J., Dantzer, R., and Kelley, K. W.,** Growth hormone augments superoxide anion secretion of human neutrophils by binding to the prolactin receptor, *J. Clin. Invest.,* 89, 451, 1992.
35. **Geffner, M. E., Bersch, N., Lippe, B. M., Rosenfeld, R. G., Hintz, R. L., and Golde, D. W.,** Growth hormone mediates the growth of T-lymphoblast cell lines via locally generated insulin-like growth factor-I, *J. Clin. Endocrinol. Metab.,* 71, 464, 1990.
36. **Guler, H. P., Zapf, J., Schweiwiller, E., and Froesch, E. R.,** Recombinant human insulin-like growth factor I stimulates growth and has distinct effects on organ size in hypophysectomized rats, *Proc. Natl. Acad. Sci. U.S.A.,* 85, 4889, 1988.
37. **Johnson, E. W., Jones, L. A., and Kozak, R. W.,** Expression and function of insulin-like growth factor receptors on anti-CD3-activated human T lymphocytes, *J. Immunol.,* 148, 63, 1992.
38. **Zelazowski, P., Dohler, K. D., Stepien, H., and Pawlikowski, M.,** Effect of growth hormone releasing hormone on human peripheral blood leukocyte chemotaxis and migration in normal subjects, *Neuroendocrinology,* 50, 236, 1989.
39. **Gupta, S., Fikrig, S. M., and Noval, M. S.,** Immunological studies in patients with isolated growth hormone deficiency, *Clin. Exp. Immunol.,* 54, 87, 1983.
40. **Kelley, K. W.,** Cross-talk between the immune and endocrine systems, *J. Anim. Sci.,* 66, 2095, 1988.
41. **Murphy, W. J., Durum, S. K., Anver, M., Frazier, M., and Longo, D. L.,** Recombinant human growth hormone promotes human lymphocyte engraftment in immunodeficient mice and results in an increased incidence of human Epstein Barr virus-induced B-cell lymphoma, *Brain, Beh. Immunol.,* 6, 355, 1992.
42. **Weigent, D. A., and Blalock, J. E.,** The effect of the administration of growth hormone-producing lymphocytes on weight gain and immune function in dwarf mice, *Neuroimmunomodulation,* 1, 50, 1994.
43. **Diamond, T., Nery, L., and Posen, S.,** Spinal and peripheral bone mineral densities in acromegaly: the effects of excess growth hormone and hypogonadism, *Ann. Intern. Med.,* 111, 567, 1989.
44. **Underwood, L. E.,** Assessment of the risk of treatment with human growth hormone, in *Proceedings of the International Symposium on Growth Hormone,* Bercu, B. B., Ed., Plenum Press, New York, 1988, 357.
45. **Mehls, O. and Tonshoff, B.,** Growth hormone in renal transplantation: the mode of action, animal studies, and clinical use, *J. Am. Soc. Nephrol.,* 2, S284, 1992.
46. **Murphy, W. J., Tsarfaty, G., and Longo, D. L.,** Growth hormone exerts hematopoietic growth-promoting effects in vivo and partially counteracts the myelosuppressive effects of azidothymidine, *Blood,* 80, 1443, 1992.

Chapter 11

SPONTANEOUS AND IATROGENIC HYPERSOMATOTROPISM

William H. Daughaday

TABLE OF CONTENTS

I. DEFINING INCREASED GROWTH HORMONE AND INSULIN-LIKE GROWTH FACTOR-I SECRETION

Hypersomatotropism is the secretion of growth hormone (GH) significantly in excess of the normal secretion. While this appears to be self-evident, in practice it is often difficult to establish. As discussed at length in Chapter 3, GH is not secreted continuously throughout the day, but is secreted in discrete secretory bursts interspersed with periods of virtual cessation of secretion.[1] In the interludes between secretion the serum GH concentration as measured with high sensitivity assays is usually <1 µg/l.[2] Much of the residual GH consists of 20 kDa GH and other variants of reduced biologic potency.[2] In order to determine the mean serum GH concentration and to estimate the amount of GH being secreted it has been necessary to sample blood at frequent intervals over a 24-period. Special methods of peak recognition and deconvolutional analysis permit estimation of GH clearance and secretory rate.[3] Because of the wide range of individual variation it may not be possible to distinguish mild GH excess from normal.

0-8493-8384-6/95/$0.00+$.50

Another complicating factor is that GH secretion varies greatly during normal development. Very high fetal serum GH concentrations normally occur during middle and late gestation.[4] These are followed by lower and relatively uniform concentration in childhood and a doubling or tripling of the concentration in late puberty.[5] After puberty GH concentrations fall so that in adults the mean serum GH concentration is between 1 and 2 µg/l with highly sensitive and specific assays. A further decline in serum GH occurs with aging.[6] In a significant number of elderly individuals the mean serum GH concentration suggests hyposomatotropism.

The interpretation of serum GH measurements is further complicated by the fact that a considerable portion of serum GH circulates associated with growth hormone binding protein (GHBP).[7] When GH is bound to GHBP it does not have equal access to target GH receptors. The concentration of GHBP is not constant throughout life. It is very low at birth and rises progressively during childhood and adolescence.[8] Adult levels of GHBP are reached in the third decade of life. Individual variability is also considerable in the concentration of GHBP.

Most of the growth-promoting actions of GH are mediated by insulin-like growth factor-I (IGF-I) and in turn serum IGF-I exerts an important negative feedback on the secretion of GH.[9] The important linkage between serum GH and IGF-I secretion varies during development. In late fetal life serum GH levels are very high but serum IGF-I concentrations are low. In childhood mean serum GH concentrations are higher than those in adult life and a progressive rise occurs in serum IGF-I. In late puberty both serum GH and IGF-I rise.[10] In adult life serum IGF-I concentrations are maintained, with very modest secretion of GH. The changing linkage between serum GH and IGF-I throughout life probably reflect increasing concentrations of hepatic GH receptors. Support for this notion is provided by the rise in serum GHBP during development. The binding protein is known to be derived from the extracellular domain of the GH receptor probably by endopeptidase cleavage.[11]

The linkage between serum GH and IGF-I in adults is well shown in the studies of acromegalic patients by Barkan et al.[12] in Figure 1. A sigmoid relationship exists between the logarithm of the mean GH concentration and the serum IGF-I concentration. When the mean serum GH rose above 3 to 4 µg/l the serum IGF-I concentration rose above normal. With further rise in serum GH there was a near-linear increase in serum IGF-I with a plateauing of serum IGF-I when serum GH reached 40 to 60 µg/l.

The linkage between serum GH concentration and IGF-I secretion is markedly influenced by nutrition.[13] Even short-term fasting results in a prompt drop in serum IGF-I. In the hypernutrition of obesity the opposite situation seems to hold with IGF-I secretion maintained despite suppressed GH secretion.[13]

The above considerations make it difficult to define hypersomatotropism solely on the basis of serum GH levels alone, and additional information from serum IGF-I measurements is required.

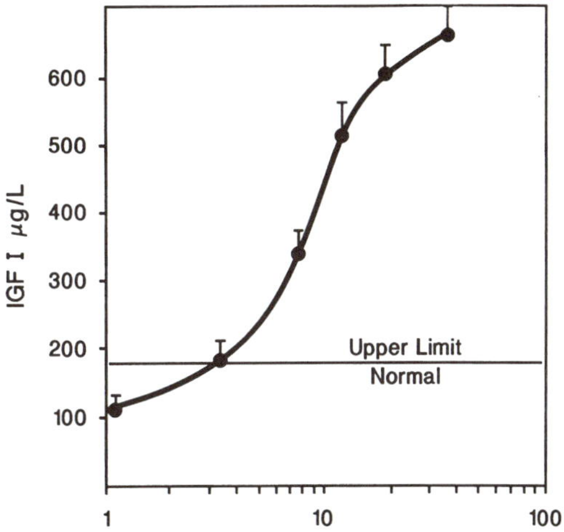

FIGURE 1. The serum IGF-I concentration was plotted against the logarithm of the mean 24-h serum GH concentration. (Reprinted from Daughaday, W. H., *Endocrinology*, 3rd ed., DeGroot, L. J., Ed., W. B. Saunders, p. 313. With permission.)

II. TYPES OF HYPERSOMATOTROPISM

Hypersomatotropism can be divided for discussion into four major types: physiologic, neoplastic, compensatory, and iatrogenic.

A. PHYSIOLOGIC GROWTH HORMONE HYPERSECRETION

The onset of puberty in both sexes begins with an increase in gonadal sex steroid secretion, which in turn acts on the hypothalamus to increase GH secretion. This action was extensively studied in hypogonadal boys.[5] There is now much evidence that testosterone must be aromatized in the hypothalamus to estrogen in order for it to stimulate GH secretion. The potentiation of GH secretion by testosterone can be blocked with a competitive inhibitor of the estrogen receptor tamoxifen.[14] In addition, dihydrotestosterone, a potent peripheral androgen which cannot be aromatized, does not promote GH secretion.[15]

The peak levels of GH and IGF-I secretion occur in late puberty at a time of maximal growth velocity.[16] It is unknown why the GH and IGF-I hypersecretion of puberty subsides after puberty is complete. Certainly there is no lack of testosterone in adult males or estrogens in menstruating women. It is fortunate that GH and IGF-I secretion do subside because if they persisted pathologic manifestations of hypersomatotropism would universally occur.

Physiologic hypersecretion of GH also occurs in pregnancy. The source of this growth hormone is the placenta and not the adenohypophysis. Placental GH is the product of the GH variant gene which is separate from the normal pituitary GH gene and is only expressed in the placenta (see Chapter 4).[17] The presence of placental GH in serum was first recognized when a monoclonal antibody which

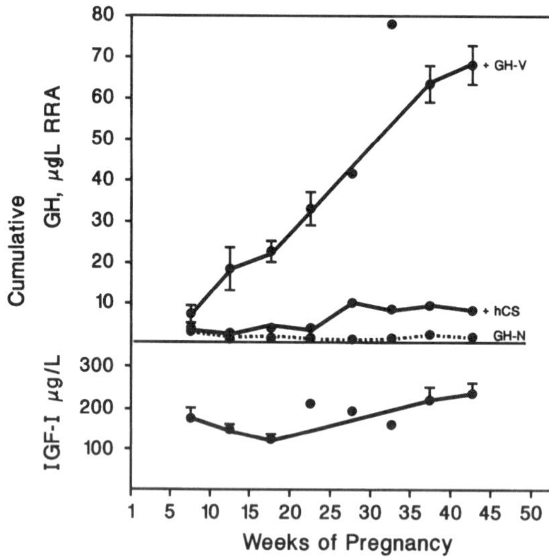

FIGURE 2. Changes in the concentration of GH-related peptides during human pregnancy. Pituitary GH was measured by a specific monoclonal antibody immunoradiometric assay. Human chorionic somatomammotropin was measured by specific immunoradiometric assay and plotted as its relative potency in the GH radioreceptor assay (0.0011). Total GH activity was measured by a human liver membrane radioreceptor assay. In pregnancy this consists almost entirely of placental GH. In the lower portion of the figure the serum IGF-I, as determined by a immunoradiometric assay, is shown. (From Daughaday, W. H., Trivedi. B., Winn, H. N., and Hong, V., *J. Clin. Endocrinol. Metab.*, 70, 215, 1990. With permission.)

reacted specifically with pituitary GH provided much lower results in serum from pregnant women than were obtained with an antibody with wider specificity. This antibody was subsequently shown to be reactive with both pituitary and placental GHs.[18,19] It was found that total GH concentrations rose progressively from the second to the third trimester and reached concentrations of 20 to 30 μg/l. When similar measurements were made with a human liver membrane radioreceptor assay the apparent total GH concentration was two- to threefold higher (Figure 2).[20] The rising titer of placental GH in serum becomes sufficient to suppress secretion of pituitary GH before the end of the first trimester. For the remainder of pregnancy virtually all the GH in serum is of placental origin. After delivery placental GH rapidly leaves the circulation and pituitary GH secretion is rapidly reestablished.

As a consequence of the GH hypersecretion during pregnancy there is a 1.5- to 3-fold increase in maternal serum IGF-I depending on the method used for measurement.[20] The increment in serum IGF-I, however, seems rather small considering the marked rise in serum GH. Evidence exists that placental GH is fully active in binding to the GH receptor and in inducing biological effects. Associated with the increased serum IGF-I there is an increase in total insulin-like growth factor binding protein-3 (IGFBP-3), but much of the IGFBP-3 is of smaller molecular size than that of nonpregnant serum as the result of placental protease action.[21] Despite the marked change in IGFBP-3 size it is still capable of entering the large molecular weight ternary complex.[22] It is not known to what extent these changes in IGFBP-3 affect free IGF-I concentrations and effectiveness of IGF-I in pregnancy serum in producing biological effects. The factors regulating the placental expression of the GH variant gene are largely unknown. Its secretation is not regulated by the hypothalamic factors regulating pituitary GH.[23]

Despite the marked changes in GH physiology which occur in human pregnancy, the placental GH is not essential for fetal development and delivery. Cases have been recognized of homozygous deletion of both the genes for placental GH and the chorionic somatomammotropin (hCS) genes.[24] Fetuses with these deletions have grown normally and have been delivered without obvious difficulty.[25] Detailed studies of pituitary GH secretion in the mothers of these fetuses have not been reported, but it may be predicted that pituitary GH would be unsuppressed and would compensate for the loss of placental GH and human chorionic somatomammotropin genes.

The hypersecretion of placental GH, and to a lesser degree hCS, in human pregnancy undoubtedly are major contributors to the insulin resistance of pregnancy and are contributing factors to the development of gestational diabetes. The hypersomatotropism of pregnancy may be the cause of the coarsening of the facial features and other mild acromegaloid changes of pregnancy.

It is unlikely that placental GH contributes to the mammary development of pregnancy because placental GH, unlike pituitary GH, has reduced affinity for the prolactin receptor.[26] Moreover, pituitary prolactin secretion, unlike that of GH, is markedly increased during pregnancy.

B. NEOPLASTIC HYPERSOMATOTROPISM

Nearly all cases of pathologic hypersomatotropism are attributable to pituitary somatotroph adenomas.[27] Adenomas can be relatively undifferentiated (stem cell adenomas) or derived from more differentiated cells which secrete both GH and prolactin (sommatomammotrophic adenomas) or only secrete GH (somatotroph adenomas). True carcinomas are very rare. Nearly all somatotroph adenomas are monoclonal and often exhibit mutations of the *Gsa* gene so that an increase occurs in adenyl cyclase activity independent of ligand occupancy of the GHRH receptor.[28–30]

Ectopic GH-secreting tumors are rare. One case of a pancreatic islet cell tumor was reported to secrete sufficient GH to cause acromegaly.[31] A duodenal carcinoid tumor in a patient with gigantism was found to contain both GH and GHRH.[32]

Approximately 1% of cases of acromegaly are the result of ectopic secretion of GHRH by an extrapituitary tumor.[33] Most cases of ectopic GHRH secretion have had either pulmonary or upper intestinal carcinoids or islet cell tumors. The pituitary pathology is varied. In some cases only somatotroph hyperplasia occurs and the sella need not be enlarged. In other cases hyperplasia progresses to nodular hyperplasia or adenoma development, usually with sellar enlargement. Very rarely hypothalamic tumors secrete GHRH, which results in GH hypersecretion.[33] The possibility of functional hypothalamic hypersecretion of GHRH occurs in rare cases of hypersomatotropism associated with somatotroph hyperplasia.

C. COMPENSATORY GROWTH HORMONE HYPERSECRETION

Whenever an impairment occurs in GH receptor binding or in postreceptor mechanisms responsible for IGF-I expression, GH hypersecretion results because reduced levels of serum IGF-I fail to exert normal negative feedback on GH secretion. The clearest example of this compensatory hypersomatotropism occurs in children with genetic defects of the GH receptor as it occurs in the GH insensitivity syndrome (Laron syndrome), discussed in Chapter 3.[34]

The ability of the liver to respond to GH with the release of IGF-I is affected early in protein-calorie malnutrition. Both GH receptor and postreceptor mechanisms were invoked to explain this functional disturbance.[35] Insulin-like growth factor-I secretion is impaired before other hepatic secretory deficiencies are evident. To compensate for this defect an increase occurs in GH secretory pulse magnitude.[36]

A similar pattern is evident in severe uncontrolled diabetes mellitus.[37,38] In this condition serum IGF-I concentrations are either low normal or low and serum GH concentrations are elevated. The GH-IGF-I axis returns to normal via control of the diabetes with insulin. Other severe medical conditions affecting hepatic, renal, or gastrointestinal function may adversely affect hepatic secretion of IGF-I. Because GH hypersecretion in these conditions is compensatory to impaired IGF-I negative feedback it would be expected that those GH responses mediated by IGF-I would not be increased, but direct actions of GH on muscle, adipose tissue, and liver would be enhanced. Also unknown is the effect of malnutrition and other serious medical conditions on the production and action of IGFs in extrahepatic tissues.

D. IATROGENIC HYPERSOMATOTROPISM

The possible therapeutic uses of GH in conditions other than childhood and adult hypopituitarism are under extensive study.[39,40] There is already evidence that GH administration can increase the growth velocity of many short children without GH deficiency. Included in these studies are children with idiopathic short stature, Turner syndrome, and growth failure secondary to chronic renal failure. In elderly adults GH treatment to increase muscle strength in the elderly, to restore bone minerals in osteoporosis, and to increase nitrogen retention in nutritional rehabilitation is under active investigation. In all these conditions GH is given in doses which are larger than endogenous secretion. If the period of treatment is prolonged the possibility exists that manifestations of hypersomatotropism might develop.

Underground use of GH in young adults to increase muscle size and strength and to increase athletic performance is widespread. This has been particularly prevalent in body builders and weightlifters despite the lack of convincing controlled experiments which show efficacy.[41]

III. MANIFESTATIONS OF HYPERSOMATOTROPISM

Most of what we know about hypersomatotropism derives from clinical studies of patients with acromegaly and gigantism. These are extremely chronic conditions and may be present for years to decades before medical attention is sought. The change in appearance of these patients is so insidious that recognition by the patient and his/her relatives or friends is usually long delayed.

A. BODY COMPOSITION

Changes in body composition occur in GH deficiency and excess. Body composition has been measured in many ways.[42] Measurements of body fat have been done by relatively simple anthropometric methods such as calculating the body mass index (weight in kilograms per height in square meters) or by measuring skinfold thickness. Recent studies use more elaborate methods, including whole body densitometry by underwater weighing or by compartmental analysis after isotopic measurement of total body water or measurement of lean body mass by measuring total body ^{42}K. The distribution of body fat in different body depots can be determined by computed tomography or magnetic resonance imaging. The measurement of bioelectrical impedance is a relatively simple, noninvasive method easily applied in the clinical setting.[43] It is based on the difference in electrical conductance of fat and fat-free mass. Considerable error can result in conditions in which a distortion of the relationship between lean body mass and interstitial fluid occurs because many formulas used in interpretation assume that this ratio is constant. Contemporary studies are increasingly applying dual energy X-ray absorptiometry.

Children with GH deficiency experience an increase in body fat and a decrease in lean body mass.[44] Following GH treatment there is a marked mobilization of body fat and an increase in lean body mass with return of body composition toward normal. Similar changes in body composition occur in hypopituitary adults with increased body fat, decreased lean body mass, and decreased extracellular water.[45] These changes return toward normal with GH treatment.

In acromegaly the changes from normal in body composition are the opposite of those noted in GH deficiency.[46] There is a decrease in body fat, an increase in lean body mass, and an increase in extracellular fluid, much of this being in interstitial fluid. The changes in body composition in acromegaly are largely reversible with successful surgical treatment.[47] Growth hormone was reported to increase Na:K pump activity.[48] In acromegaly total body potassium in increased. Some of this increase is attributable to a relative increase in lean body mass, but there is also an increase in muscle potassium concentration which was attributed to increased activity of the Na:K pump.[49] The increase in extracellular fluid volume in acromegaly was also attributed to increased activity of the Na:K pump by shifting muscle sodium to the extracellular fluid. This explanation is unlikely because there is insufficient muscle sodium to maintain an increase in extracellular water. Moreover, the volume and electrolyte concentration of the extracellular fluid are independently regulated. It is much more likely that the major increment in extracellular water is in the interstitial compartment resulting from the deposition of hydrophilic proteoglycans in the connective tissue and skin. The increase in interstitial water results in characteristic nonpitting edema of acromegaly, which is most noticeable clinically in facial features and acral parts. It is likely that these changes in interstitial water are mediated by IGF-I because evidence of interstitial edema was observed in subjects receiving IGF-I for several months.[50] The facial and acral puffiness of acromegaly is rapidly reversible after successful removal of a somatotroph adenoma. Marked improvement in appearance can be noted within 1 week. This is possible because the half-life of hyaluronate in connective tissue is only 2 to 3 weeks.

B. ALTERED INTERMEDIARY METABOLISM

Growth hormone has an immediate but transient insulin-like action on blood glucose concentration. The major action of GH on carbohydrate metabolism is to reduce the ability of insulin to increase uptake of glucose into muscle and to inhibit the release of glucose from the liver. To compensate an increase takes place in both the fasting and postprandial serum insulin concentrations in acromegaly.[51] The β cells of the pancreatic islets respond to this increase in insulin demand by increasing insulin synthesis and in younger animals by β cell hyperplasia. In addition GH may have a direct islet-tropic action, as has been shown *in vitro*. In most younger individuals increased insulin secretion is usually sufficient to maintain euglycemia, but in some older patients with acromegaly, particularly those with a family history of diabetes, impaired glucose tolerance or frank diabetes develops.

The insulin antagonism of acromegaly is a direct action of GH and is not IGF-I mediated. Insulin-like growth factor-I has a truly insulin-like action on muscle and spares insulin.[52] Growth hormone can act through its receptors in muscle, adipocytes, and hepatocytes. In muscle insulin insensitivity may be the result of decrease in mobilization of the Glut-4 glucose transporter.

Growth hormone has a major action in increasing protein synthesis in many tissues. This was shown with measurements of nitrogen balance and with studies of ^{3}H-leucine disappearance from the circulatory pool.[53] These studies show that after GH treatment of hypopituitary patients the flux of amino acids into the nonoxidative pathway of protein synthesis is increased. It appears that the direct action of GH on tissue protein synthesis is more important than an indirect action through IGF-I. This may be inferred from the comparison of the effects of GH administration on nitrogen balance with those obtained when supraphysiologic doses of IGF-I are given. A small role for IGF-I in total body nitrogen metabolism was evident in the studies of Clemmons and Underwood[54] of malnourished patients. There was an additive effect on nitrogen retention when IGF-I administration was combined with GH administration.

As the measurements of body composition revealed, GH has a major effect on fat metabolism.[55] Growth hormone increases lipolysis in fat depots, increases plasma free fatty acid concentration, and increases the fraction of total energy derived from fat oxidation and hepatic ketogenesis. The action of GH on fat metabolism is direct and does not require the intermediary action of IGF-I. Both human adipocytes and hepatocytes have few IGF-I receptors.

C. CHANGES IN SKIN AND CONNECTIVE TISSUES

In GH excess there is thickening of the skin and stimulation of growth and function of skin appendages. This can be quantified by measurement of the skin-fold thickness on the back of the hand or the thickness of the heel pad as seen on X-ray films. Hair follicles are stimulated and mild hypertrichosis occurs in women. Sweat and sebaceous glands hypertrophy and patients frequently complain of copious, odoriferous sweat. Skin tags (acrochordons) are more numerous and may drop off after successful treatment. In longstanding acromegaly the scalp may heap up in horizontal ridges.

Connective tissue proliferation is not limited to the skin and, as previously mentioned, results in an expansion in interstitial water. Connective tissue proliferation is largely responsible for macroglossia, which can deform the mandible and be a factor in sleep apnea, a condition which is particularly prevalent in acromegaly.[56] Larygeal thickening in long-established acromegaly lowers the pitch of the voice. Connective tissue swelling within the limited confines beneath the carpal ligament results in median nerve dysfunction.[57] Entrapment of the posterior tibial nerve by the tarsal ligament also occurs.

D. CHANGES IN BONES AND JOINTS

Excessive skeletal growth is the principal manifestation of hypersomatotropism that develops before the closure of the epiphyseal growth plates, resulting in gigantism.[58] After puberty, growth in the length of the bones is limited to the mandible where proliferation of cartilage at the temporomandibular joint occurs and results in prognathism. Limited growth in length of the ribs can occur by proliferation of cartilage at the costal chondral junctions resulting in a barrel chest deformity.

Bone metabolism is changed in acromegaly.[59] Serum calcium level is usually normal but serum phosphorus concentration is elevated by a direct action of GH on renal tubular reabsorption of phosphorus. Urinary calcium excretion is increased in about a 25% of the patients. Despite these minor changes in mineral metabolism, there is usually a marked increase in indices of bone turnover, including the urinary hydroxyproline:creatinine ratio, urinary type I collagen crosslinked *N*-telepeptide, and serum osteocalcin. Early in acromegaly there may be an increase in trabecular bone density, but this has been hard to document. The usual finding in established acromegaly is a decrease in vertebral trabecular bone density.[59] Vertebral osteopenia and kyphosis are common.[60] The hypogonadism that is frequent in longstanding acromegaly may be important in the loss of bone density.

Articular cartilage proliferation continues in acromegaly and leads to widening of the joint space, which may be recognized by X-ray study. The transfer of nutrients through the thickened cartilage is impaired and joint mechanics are frequently altered, resulting in painful degenerative arthritis.

E. VISCERAL SIZE AND FUNCTION

The kidneys respond to chronic hypersomatotropism with an in increase in size and function. Both glomerular filtration and renal plasma flow are increased. These responses are probably mediated by IGF-I reaching the kidney as an endocrine factor and the local production of IGF-II in the kidney acting in an autocrine/paracrine fashion.[61] Type I IGF receptors are widely distributed in the kidney, with the greatest concentration in the tubulo-interstitial region. Insulin-like growth factors play an intermediate role in GH action on the kidney as shown by the finding that IGF-I administered to normal volunteers increases glomerular filtration.[62] Despite years of glomerular hyperfiltration, there is little evidence that this leads to glomerular sclerosis unless diabetes supervenes.

The weight of the heart is increased in acromegaly, with thickening both of the left ventricular wall and septum.[63,64] The former may limit cardiac filling and the latter may impede ejection into the aorta. There is little evidence that atherosclerosis is accelerated. The cardiac changes plus hypertension may account for the increase in cardiovascular mortality.

Liver weight is increased in acromegaly and there is enhanced biliary transport as well as other functions. There appears to be no long-term detrimental effects of hypersomatotropism on the liver.

F. NEUROLOGIC COMPLICATIONS

Pituitary tumors with suprasellar extension frequently impinging on the optic chiasm to produce visual field defects. Tumor extension can rarely cause other cranial nerve involvement. Upward extension into the hypothalamus can produce hydrocephalus and a variety of hypothalamic consequences.

The association of acromegaly with sleep apnea and peripheral nerve entrapment were already mentioned.[56,57] A form of diffuse hypertrophic neuropathy occurs in gigantism and in some cases of acromegaly and may be disabling.[65,66]

Muscle strength may be increased early in acromegaly, but later in the course of the disease weakness is more common. In some cases this is attributable to neuropathy, but an acromegalic myopathy has been described.[65]

G. ASSOCIATED MALIGNANCIES

Much interest has been generated over the years in the possible link between hypersomatotropism with the occurrence or progression of malignancies.[67] Rats given GH continuously for up to 485d suffered from a variety of tumors, including pulmonary lymphosarcoma, adrenal cortical and medullary tumors, and tumors of the breast and ovary.[68] These tumors are not rare in older rats, but their frequency appeared to be increased by GH.

Many tumor cell lines grow *in vitro* without the addition of growth factors. In some cases growth can be prevented by antibodies directed at IGF-I or the IGF-I receptors.[69] These cells can be shown to express the IGF-I gene, which can act as an autocrine/paracrine growth factor.

It has been harder to establish a relationship between GH excess and an increased risk of malignancies in people with acromegaly. Two large retrospective surveys failed to find an increased risk of malignant tumors in acromegaly.[70,71] In a more recent follow-up of 166 acromegalic patients a threefold increase in death occurred due to malignancies compared to a population-based control group.[72] There is evidence that colon polyps and cancer are more common in acromegaly.[67] In a small prospective study of acromegalic patients, 53% were found to harbor polyps.[73] The prevalence of colon cancer among acromegalic patients has been reported to be 6.9%.[74] An association has been recognized between colonic polyps and multiple skin acrochordons. The presence of more than three of these tags increases the likelihood of finding colonic polyps.

IV. COMPLICATIONS OF GROWTH HORMONE TREATMENT

There have been few significant complications of GH treatment. The vast majority of patients treated have suffered from pituitary deficiency and the doses of GH used have been constrained by supply or cost. The most devastating complication of treatment with pituitary-derived human GH the development of fatal encephalopathy (Creutzfeldt-Jakob disease).[75] A total of 28 cases of this untreatable condition have been recognized, among >8000 children at risk. As soon as this complication was recognized, GH prepared from human pituitaries was withdrawn from human use in the U.S., and soon was replaced by recombinant GH.

Although as many as 5 to 10% of children receiving recombinant GH develop low titer antibodies directed against GH, these rarely reach titers which seriously impair the effectiveness of administered GH. High-titer antibodies that block GH effects develop after an initial favorable growth response in children with deletions or other mutations of the GH gene which prevents secretion of any GH.[76] Such patients are candidates for treatment with IGF-I. At first the concern was that GH administration to normal experimental subjects might provoke antibody responses which could inhibit endogenous GH action. This did not occur.

About the only manifestation of hypersomatotropism that has been recognized in patients receiving therapeutic GH has been fluid retention. Lesser degrees of fluid retention probably are common, but only rarely does it become sufficient to be recognized clinically. Patients with Turner syndrome are said to be more frequently affected and this is attributed to inadequate lymphatic drainage, which frequently leads to neonatal peripheral edema.[77] An equally likely explanation is that patients with Turner syndrome often receive somewhat higher doses of GH.

The fluid which is retained early after the institution of GH therapy is attributable in part to stimulation of the renin-angiotensin-aldosterone system, which increases renal reabsorption of sodium.[78] More persistent nonpitting puffiness of the facial features and extremities is the result of interstitial fluid space expansion similar to that which occurs in acromegaly.[79] It is likely that this interstitial edema is IGF-I mediated because it also occurs in adults treated with IGF-I over a prolonged period.[80] Children with this mild acromegaloid facial swelling during GH treatment have been found to have elevated serum IGF-I concentrations.

A number of other complaints occur in patients receiving either GH or IGF-I which may relate to interstitial edema. These include carpal-tunnel syndrome, parotid swelling, and facial discomfort. Pseudotumor cerebri was reported in rare patients receiving GH, but its pathogenesis is obscure.

The treatment of GH-deficient children with replacement GH has seldom impaired glucose tolerance. Basal and postprandial serum insulin concentrations increase, but these changes restore normal patterns of insulin secretion. Aging is associated with a progressive decrease in insulin secretory capacity, and this is marked in patients with the genetic determinants for noninsulin-dependent diabetes. The development of glucose intolerance and even diabetes has followed GH administration to older adults.[81] This may prove to be a significant problem if GH use becomes established in geriatric medicine.

Excessive skeletal growth is an unlikely consequence of legitimate GH treatment, but this may be encountered in the future if GH is used indiscriminately to increase stature or to enhance athletic performance.

Slipped capital femoral epiphysis occurs more frequently during periods of rapid growth. Its occurrence in patients with pituitary deficiency who are receiving GH is probably the consequence of rapid growth rather than evidence of GH toxicity.[82]

A cluster of ten cases of leukemia occurred in GH-treated patients in Japan. This raised considerable concern in the pediatric endocrine community and among parents of children receiving GH. Careful follow-up studies in the U.S., Canada, U.K., and Europe have not found any association between GH treatment and the incidence of leukemia.[83] Cerebral irradiation required in the treatment of acute leukemia in children frequently results in partial GH deficiency. Subsequent GH treatment was not found to increase the risk of recurrent cerebral or systemic leukemia in such patients.[84]

V. CONCLUSION

In summary, spontaneous hypersomatotropism is a serious medical condition leading to gigantism and acromegaly. This can lead to disfigurement, diabetes mellitus, increase in cardiovascular mortality, neurologic complications, and possible increased risk of malignancies. Extensive experience with the use of recombinant human GH in the treatment of pituitary deficiency has indicated that manifestations of hypersomatotropism are seldom encountered. New indications for GH treatment are being explored. It is likely that if adults are to be treated, especially if they receive supraphysiologic doses for extended periods, adverse changes of hypersomatotropism will be encountered.

REFERENCES

1. **Zadik, Z., Chalew, S. A., McCarter, R. J., Jr., Meistas, M., and Kowarski, A. A.,** The influence of age on the 24-hour integrated concentration of growth hormone in normal individuals, *J. Clin. Endocrinol. Metab.*, 60, 513, 1985.
2. **Baumann, G.,** Growth hormone heterogeneity: genes, isohormones, variants and binding proteins, *Endocr. Rev.*, 12, 424, 1991.
3. **Veldhuis, J. D. and Johnson, M. L.,** Deconvolutional analysis of hormone data methods, *Methods Enzymol.*, 210, 539, 1992.
4. **Gluckman, P. D., Grumbach, M. M., and Kaplan, S. L.,** The neuroendocrine regulation and function of growth hormone and prolactin in the mammalian fetus, *Endocr. Rev.*, 2, 363, 1981.
5. **Rogol, A. D.,** Growth and growth hormone secretion at puberty: the role of gonadal steroid hormones, *Acta Paediatr. (Suppl.)*, 383, 15, 1992.
6. **Corpas, E., Harman, S. M., and Blackman, M. R.,** Human growth hormone and human aging, *Endocr. Rev.*, 14, 20, 1993.
7. **Martha, P. M., Jr., Reiter, E. O., Davila, N., Shaw, M. A., Holcombe, J. H., and Baumann, G.,** Serum growth hormone (GH)-binding protein/receptor: an important determinant of GH responsiveness, *J. Clin. Endocrinol. Metab.*, 75, 1464, 1992.
8. **Merimee, T. J., Russell, B., and Quinn, S.,** Growth hormone-binding proteins of human serum: developmental patterns in normal man, *J. Clin. Endocrinol. Metab.*, 75, 852, 1992.
9. **Tannenbaum, G. S. and Ling, N.,** The interrelationship of growth hormone (GH)-releasing factor and somatostatin in generation of the ultradian rhythm of GH secretion, *Endocrinology*, 115, 1952, 1984.
10. **Rosenfeld, R. G., Wilson, D. M., Lee, P. D. K., and Hintz, R. L.,** Insulin-like growth factors I and II in evaluation of growth retardation, *J. Pediatr.*, 109, 428, 1986.
11. **Leung, D. W., Spencer, S. A., Cachianes, G., Hammonds, R. G., Collins, C., Henzel, W. J., Barnard, R., Waters, M. J., and Wood, W. I.,** Growth hormone receptor and serum binding protein: purification, cloning and expression, *Nature*, 330, 537, 1987.
12. **Barkan, A. L., Beitins, I. Z., and Kelch, R. P.,** Plasma insulin-like growth factor-I/somatomedin-C in acromegaly: correlation with the degree of growth hormone hypersecretion, *J. Clin. Endocrinol. Metab.*, 67, 69, 1988.
13. **Vance, M. L., and Hartman, M. L., and Thorner, M. O.,** Growth hormone and nutrition, *Horm. Res.*, 38(Suppl.), 1, 85, 1992.
14. **Weissberger, A. J., and Ho, K. K. Y.,** Activation of the somatotropic axis by testosterone in adult males: evidence for the role of aromatization, *J. Clin. Endocrinol. Metab.*, 76, 1407, 1993.
15. **Keenan, B. S., Richards, G. E., Ponder, S. W., Dallas, J. S., Nagamani, M., and Smith, E. R.,** Androgen-stimulated pubertal growth: the effects of testosterone and dihydrotestosterone on growth hormone and insulin-like-growth factor-I in the treatment of short stature and delayed puberty, *J. Clin. Endocrinol. Metab.*, 76, 996, 1993.
16. **Cara, J. F., Rosenfield, R. L., and Furlanetto, R. W.,** A longitudinal study of the relationship of plasma somatomedin-C concentration to the pubertal growth spurt, *Am. J. Dis. Child.*, 141, 562, 1987.
17. **Frankenne, F., Rentier-Delrue, F., Scippo, M.-L., Martial, J., and Hennen, G.,** Expression of the growth hormone variant gene in human placenta, *J. Clin. Endocrinol. Metab.*, 64, 635, 1987.

18. **Hennen, G., Frankenne, F., Closset, J., Gomez, F., Pirens, G., and El Khayat, N.,** A human placental GH: increasing levels during second half of pregnancy with pituitary GH suppression as revealed by monoclonal antibody radioimmunoassays, *Int. J. Fertil.*, 30(2), 27, 1985.
19. **Caufriez, A., Frankenne, F., Englert, Y., Golstein, J., Cantraine, F., Hennen, G., and Copinschi, G.,** Placental growth hormone as a potential regulator of maternal IGF-I during human pregnancy, *Am. J. Physiol.*, 258, E1014, 1990.
20. **Daughaday, W. H., Trivedi, B., Winn, H. N., and Yan, H.,** Hypersomatotropism in pregnant women, as measured by a human liver radioreceptor assay, *J. Clin. Endocrinol. Metab.*, 70, 215, 1990.
21. **Binoux, M., Hossenlopp, P., Lassare, C., and Segovia, B.,** Degradation of IGF binding protein-3 by proteases: physiological implications, in *Modern Concepts of Insulin-Like Growth Factors*, Spencer, E. M., Ed., Elsevier, Amsterdam, 1991, 329.
22. **Gargosky, S. E., Owens, P. C., Walton, P. E., Owens, J. A., Robinson, J. S., Wallace, J. C., and Ballard, F. J.,** Most of the circulating insulin-like growth factors-I and -II are present in the 150 kDa complex during human pregnancy, *J. Endocrinol.*, 131, 491, 1991.
23. **De Zegher, F., Vanderschueren-Lodeweyckx, M., Spitz, B., Faijerson, Y., Blomberg, F., Beckers, A., Hennen, G., and Frankenne, F.,** Perinatal growth hormone (GH) physiology: effect of GH-releasing factor on maternal and fetal secretion of pituitary and placental GH, *J. Clin. Endocrinol. Metab.*, 71, 520, 1990.
24. **Wohlk, P., Nexo, E., Jorgensen, E. H., Chemnitz, J., Nielsen, P. V., and Parks, J. S.,** Low or absent serum placental lactogen hormone in two normal pregnancies, *Ugeskr. Laeger*, 146, 727, 1984.
25. **Wurzel, J. M., Parks, J. S., Herd, J. E., and Nielsen, P. V.,** Gene deletion is responsible for absence of immunoassayable human chorionic somatomammotropin, *DNA* 1, 251, 1982.
26. **Ray, J., Okamura, H., Kelly, P. A., Cooke, N. E., and Liebhaber, S. A.,** Human growth hormone-variant demonstrates a receptor binding profile distinct from that of normal pituitary growth horomone, *J. Biol. Chem.*, 265, 7939, 1990.
27. **Asa, S. L. and Kovacs, K.,** Pituitary pathology in acromegaly, *Endocrinol. Metab. Clin. North Am.*, 21, 553, 1992.
28. **Herman, V., Fagin, J., Gonsky, R., Kovacs, K., and Melmed, S.,** Clonal origin of pituitary adenomas, *J. Clin. Endocrinol. Metab.*, 71, 1427, 1990.
29. **Melmed, S.,** Etiology of pituitary acromegaly, *Endocrinol. Metab. Clin. North Am.*, 21, 539, 1992.
30. **Landis, C. A., Masters, S. B., Spada, A., Pace, A. M., Bourne, H. R., and Vallar, L.,** GTPase inhibiting mutations activate the α chain of G_s and stimulate adenylyl cyclase in human pituitary tumors, *Nature*, 340, 692, 1989.
31. **Melmed, S., Ezrin, C., Kovacs, K., Goodman, R. S., and Frohman, L. A.,** Acromegaly due to secretion of growth hormone by an ectopic pancreatic islet-cell tumours, *N. Engl. J. Med.*, 312, 9, 1985.
32. **Leveston, S. A., McKeel, D. W., Jr., Buckley, P. J., Deschryver, K., Greider, M. H., Jaffe, B. M., and Daughaday, W. H.,** Acromegaly and Cushing's syndrome associated with a foregut carcinoid tumor, *J. Clin. Endocrinol. Metab.*, 52, 682, 1981.
33. **Kovaks, K.,** Growth hormone-releasing hormone-producing tumors: clinical, biochemical, and morphological manifestations, *Endocr. Rev.*, 9, 357, 1987.
34. **Laron, Z., Pertzelan, A., Karp, M., Keret, R., Eshet, R., and Silbergeld, A.,** Laron syndrome—a unique model of IGF-I deficiency, in *Lessons from Laron Syndrome (LS)*, Pediatr. Adolesc. Endocrinol. Vol. 24, Laron, Z., and Parks, J. S., Eds., S. Karger, Basel, 1993, 3.
35. **Straus, D. S., and Takemoto, C. D.,** Effect of fasting on insulin-like growth factor-I (IGF-I) and growth hormone receptor mRNA levels and IGF-I gene transcription in rat liver, *Mol. Endocrinol.*, 4, 91, 1990.

36. **Hartman, M. L., Veldhuis, J. D., Johnson, M. L., Lee, M. M., Alberti, K. G. M. M., Samojlik, E., and Thorner, M. O.,** Augmented growth hormone (GH) secretory burst frequency and amplitude mediate enhanced GH secretion during a two-day fast in normal men, *J. Clin. Endocrinol. Metab.,* 74, 757, 1992.
37. **Horner, J. M., Kemp, S. F., and Hintz, R. L.,** Growth hormone and somatomedin in insulin-dependent diabetes mellitus, *J. Clin. Endocrinol. Metab.,* 53, 1148, 1981.
38. **Rieu, M., and Binoux, M.,** Serum levels of insulin-like growth factor (IGF) and IGF binding protein in insulin-dependent diabetics during an episode of severe metabolic decompensation and the recovery phase, *J. Clin. Endocrinol. Metab.,* 60, 781, 1985.
39. **Lippe, B. M., and Nakamoto, J. M.,** Conventional and nonconventional uses of growth hormone, *Rec. Prog. Horm. Res.,* 48, 179, 1993.
40. **Westphal, O.,** Non-coventional growth hormone treatment in short children, *Acta Endocrinol.,* 128(Suppl. 2), 10, 1993.
41. **Deyssig, R., Frisch, H., Blum, W. F., and Waldhor, T.,** Effect of growth hormone treatment on hormonal parameters, body composition and strength in athletes, *Acta Endocrinol.,* 128, 313, 1993.
42. **Yarasheski, K. E., Zachwieja, J. J., Angelopoulos, T. J., and Bier, D. M.,** Short-term growth hormone treatment does not increase muscle protein synthesis in experienced weight lifters, *J. Appl. Physiol.,* 74, 3073, 1993.
43. **Brummer, R.-J. M, Rosen, T., and Bengtsson, B.-A.,** Evaulation of different methods of determining body composition with special reference to growth hormone-related disorders, *Acta Endocrinol.,* 128(Suppl. 2), 30, 1993.
44. **Collipp, P. J., Curti, V., Thomas, J., Sharma, R. K., Maddaiah, V. T., and Cohn, S. H.,** Body composition changes in children receiving human growth hormone, *Metab. Clin. Exp.,* 22, 589, 1973.
45. **Bengtsson, B. Å., Edén, S., Lonn, L., Kvist, H., Stokland, A., Lindstedt, G., Bosaeus, I., Tolli, J., Sjostrom, L., and Isaksson, O. G. P.,** Treatment of adults with growth hormone (GH) deficiency with recombinant human GH, *J. Clin. Endocrinol. Metab.,* 76, 309, 1993.
46. **Bengtsson, B. Å., Brummer, R.-J. M., Edén, S., and Bosaeus, I.,** Body composition in acromegaly, *Clin. Endocrinol.,* 30, 121, 1989.
47. **Bengtsson, B. Å., Brummer, R.-J., Edén, S., Bosaeus, I., and Lindstedt, G.,** Body composition in acromegaly: the effect of treatment, *Clin. Endocrinol.,* 31, 481, 1989.
48. **Shimomura, Y., Lee, M., Oku, J., Bray, G. A., and Glick, Z.,** Sodium potassium dependent ATPase in hypophysectomized rats: response to growth hormone, triiodothyronine and cortisone, *Metab. Clin. Exp.,* 31, 213, 1982.
49. **Landin, K., Petruson, B., Jakobsson, K. E., and Bengtsson, B. A.,** Skeletal muscle sodium and potassium changes after successful surgery in acromegaly: relation to body composition, blood glucose, plasma insulin and blood pressue, *Acta Endocrinol.,* 128, 418, 1993.
50. **Underwood, L. E., Backeljauw, P., and Clemmons, D. R.,** Actions of IGF-I in vivo, in *Proceedings of the Third International Symposium on Insulin-Like Growth Factors,* Baxter, R. C., Gluckman, P. D., and Rosenfeld, R. G., Eds., Elsevier, New York, in press.
51. **Cerasi, E., Luft, R.,** Insulin response to glucose leading in acromegaly, *Lancet,* 2, 769, 1964.
52. **Guler, H.-P., Zapf, J., and Froesch, E. R.,** Short-term metabolic effects of recombinant human insulin-like growth factor I in healthy adults, *N. Engl. J. Med.,* 317, 137, 1987.
53. **Russell-Jones, D. L., Weissberger, A. J., Bowes, S. B., Kelly, J. M., Thomason, M., Umpleby, A. M., Jones, R. H., and Sonksen, P. H.,** The effects of growth hormone on protein metabolism in adult growth hormone deficient patients, *Clin. Endocrinol.,* 38, 427, 1993.
54. **Clemmons, D. R., and Underwood, L. E.,** Role of insulin-like growth factors and growth hormone in reversing catabolic states, *Horm. Res.,* 38(Suppl.), 2, 37, 1992.
55. **Goodman, H. M., Schwartz, Y., Tai, L. R., and Gorin, E.,** Actions of growth hormone on adipose tissue: possible involvement of autocrine and paracrine factors, *Acta Paediatr. Scand.* (Suppl.), 367, 132, 1990.

56. **Trotman-Dickenson, B., Weetman, A. P., and Hughes, J. M. B.,** Upper airway obstruction and pulmonary function in acromegaly: relationship to disease activity, *Q. J. Med.*, 79(NS), 527, 1991.
57. **O'Duffy, J. D., Randall, R. V., and MacCarty, C. S.,** Median neuropathy (carpal-tunnel syndrome) in acromegaly. A sign of endocrine overactivity, *Ann. Intern. Med.*, 78, 379, 1973.
58. **Daughaday, W. H.,** Pituitary gigantism, *Endocrinol. Metab. Clin. North Am.*, 21, 633, 1992.
59. **Ezzat, S., Melmed, S., Endres, D., Eyre, D. R., and Singer, F. R.,** Biochemical assessment of bone formation and resorption in acromegaly, *J. Clin. Endocrinol. Metab.*, 76, 1452, 1993.
60. **Whitehead, E. M., Shalet, S. M., Davies, D., Enoch, B. A., Price, D. A., and Beardwell, C. G.,** Pituitary gigantism: a disabling condition, *Clin. Endocrinol.*, 17, 271, 1982.
61. **Chin, E. and Bondy, C.,** Insulin-like growth factor system gene expression in the human kidney, *J. Clin. Endocrinol. Metab.*, 75, 962, 1992.
62. **Guler, H. P., Schmid, C., Zapf, J., and Froesch, E. R.,** Effects of recombinant insulin-like growth factor I on insulin secretion and renal function in normal human subjects, *Proc. Natl. Acad. Sci. U.S.A.*, 86, 2868, 1989.
63. **Smallridge, R. C., Rajfer, S., Davia, J., and Schaaf, M.,** Acromegaly and the heart. An echocardiographic study, *Am. J. Med.*, 66, 22, 1979.
64. **Hayward, R. P., Emanuel, R. W., and Nabarro, J. D. N.,** Acromegalic heart disease: influence of treatment of the acromegaly on the heart, *Q. J. Med.*, 62, 41, 1987.
65. **Pickett, J. B. E., III, Layzer, R. B., Levin, S. R., Schneider, V., Campbell, M. J., and Summer, A. J.,** Neuromuscular complications of acromegaly: relationship to disease activity, *Neurology*, 25, 638, 1975.
66. **Daughaday, W. H.,** Extreme gigantism, *N. Engl. J. Med.*, 297, 1267, 1977.
67. **Ezzat, S. and Melmed, S.,** Are patients with acromegaly at increased risk for neoplasia, *J. Clin. Endocrinol. Metab.*, 72, 245, 1991.
68. **Moon, H. D., Simpson, M. E., Li, C. H., and Evans, H. M.,** Neoplasms in rats treated with pituitary growth hormone I. Pulmonary and lymphatic tissues, *Cancer Res.*, 10, 297, 1950.
69. **Daughaday, W. H. and Deuel, T. F.,** Tumor secretion of growth factors, *Endocrinol. Metab. Clin. North Am.*, 20, 539, 1991.
70. **Wright, A. D., Hill, D. M., Lowy, C., and Fraser, T. R.,** Mortality in acromegaly, *Q. J. Med.*, 39, 1, 1970.
71. **Alexander, L., Appleton, D., Hall, R., Ross, W. M., and Wilkinson, R.,** Epidemiology of acromegaly in the Newcastle region, *Clin. Endocrinol.*, 12, 71, 1980.
72. **Bengtsson, B. Å., Edén, S., Ernest, I., Odén, A., and Sjogren, B.,** Epidemiology and long-term survival in acromegaly, *Acta Med. Scand.*, 223, 327, 1988.
73. **Klein, I., Parveen, G., Gavaler, J. S., and Vanthiel, D. H.,** Colonic polyps in patients with acromegaly, *Ann. Intern. Med.*, 97, 27, 1982.
74. **Brunner, J. E., Johnson, C. C., Zafar, S., Peterson, E. L., Brunner, J. F., and Mellinger, R. C.,** Colon cancer and polyps in acromegaly: increased risk associated with family history of colon cancer, *Clin. Endocrinol.*, 32, 65, 1990.
75. **Fradkin, J. E., Schonberger, L. B., Mills, J. L., Gunn, W. J., Piper, J. M., Wysowski, D. K., Thomson, R., Durako, S., and Brown, P.,** Creutzfeldt-Jakob disease in pituitary growth hormone recipients in the United States, *JAMA*, 265, 880, 1991.
76. **Phillips, J. A., III and Cogan, J. D., III,** Genetic basis of endocrine disease. VI. Molecular basis of familial human growth hormone deficiency, *J. Clin. Endocrinol. Metab.*, 78, 11, 1994.
77. **Hintz, R. L.,** Untoward events in patients treated with growth hormone in the USA, *Horm. Res.*, 38(Suppl.), 1, 44, 1992.
78. **Ho, K. Y. and Weissberger, A. J.,** The antinatriuretic action of biosynthetic human growth hormone involves activation of the renin-angiotensin system, *Metabolism*, 39, 133, 1990.
79. **Baens-Bailen, R., Foley, T. P., Hintz, R. L., and Lee, P. A.,** Excessive growth hormone dosing in GH deficiency, *Pediatr. Res.*, 31, 731, 1992.

80. **Underwood, L. E., Backeljauw, P., and Clemmons, D. R.,** Actions of IGF-I in vivo, in *Proceedings of the Third International Symposium on Insulin-Like Growth Factors,* Baxter, R. C., Gluckman, P. D., and Rosenfeld, R. G., Eds., Elsevier, New York, in press.
81. **Marcus, R., Butterfield, G., Holloway, L., Gilliland, L., Baylink, D. J., Hintz, R. L., and Sherman, B. N.,** Effects of short term administration of recombinant human growth hormone to elderly people, *J. Clin. Endocrinol. Metab.,* 70, 519, 1990.
82. **Rappaport, E. B. and Fife, D.,** Slipped capital femoral epiphysis in growth hormone-deficient patients, *Am. J. Dis. Child.,* 136, 396, 1985.
83. **Stahnke, N. and Zeisel, H. J.,** Growth hormone therapy and leukemia, *Eur. J. Pediatr.,* 143, 591, 1989.
84. **Ogilvy-Stuart, A., Ryder, W. D., Gattamaneni, H. R., Clayton, P. E., and Shalet, S. M.,** Growth hormone and tumour recurrence, *Br. Med. J.,* 304, 1601, 1992.

INDEX

M

N

O

P

Q

R

S